SUDOKU FOR BRAIN FITNESS

90-DAY CHALLENGE
to Sharpen the Mind and Strengthen Cognitive Skills

STEVEN CLONTZ AND **JULIE DEMYANOVICH**

ZEITGEIST • NEW YORK

Published in the United States by Zeitgeist, an imprint of Zeitgeist™,
a division of Penguin Random House LLC, New York.
zeitgeistpublishing.com

ISBN: 978-0-593-68981-3

Book design by Paul Barrett and Rachel Marek
Edited by Ada Fung
Introduction written by L. J. Tracosas

Printed in the United States of America
1st Printing

CONTENTS

INTRODUCTION
Ready, Set, Solve!

Welcome to the 90-Day Sudoku Challenge! Whether you're a sudoku novice or you already play regularly, you'll find that making a daily practice of solving sudoku puzzles is more than just a fun way to pass time—you can also reap potential brain benefits. In this introduction, we'll walk you through those benefits, offer some puzzle-solving techniques, and explain how the 90-day challenge works and what you can gain from completing it.

THE BRAIN BENEFITS OF DAILY SUDOKU

Just like the rest of your body, your brain needs exercise and care to function at its best. The good news is, exercising your brain doesn't have to feel like a chore.

Completing puzzles like sudoku provides an excellent workout for your brain.

Research suggests that there's a correlation between regularly playing games like sudoku and improving one's memory and ability to focus. For example, in one study by the University of Exeter and King's College London published in the *International Journal of Geriatric Psychiatry*, participants who played word and puzzle games every day performed as well as people *eight years younger* than them on cognition tests.

While more research is needed to definitively corroborate these findings, many experts agree that daily puzzling challenges your mind. When working on a sudoku grid, you use problem-solving and decision-making skills; rely on working memory, logic, and analysis; and improve your powers of concentration. Sudoku is like a full-body strength class for your brain!

As with any other form of exercise, it's best to do at least a little bit every day to keep your brain fit and functioning at its highest levels. By completing puzzles each day, you're also dedicating time to your well-being. Studies show that, generally, it takes about 66 days to form a habit (though that number can vary depending on the person and the action). Thus, throughout the 90-Day Sudoku Challenge, you'll not only give your brain a workout but also build a habit of self-care that you can stick to.

SUDOKU SOLVING TIPS AND TECHNIQUES

Sudoku is a gridded, fill-in-the-blank puzzle. Grids are nine cells wide and nine cells tall. Each large grid is divided into nine smaller, 3-by-3 grids. Every 3-by-3 grid, 9-cell column, and 9-cell row must contain the numbers 1 through 9, with no repetition. All puzzles contain a few givens—that is, cells that already have a number filled in. Use the givens and the rules to fill in the empty cells.

While easy sudoku puzzles may have simple, solve-at-a-glance answers to some cells, more challenging puzzles have few to no obvious clues. For those, you'll need to rely on solving techniques. Here are a few that might help.

> **TIP: Work in pencil!** So you can erase mistakes and use notes or pencil marks—a strategy outlined in technique #4 and relied on by sudoku puzzle pros.

#1. Spot single empty cells. Look for easy "gimmes": These are the single empty cells in an otherwise complete 3-by-3 grid, row, or column. Since numbers can't repeat in a given grid, row, or column, filling in a single empty cell will be a breeze.

#2. Scan the grid. For this technique, simply sweep your eyes over the puzzle and analyze the information it gives you. Start by focusing on one 3-by-3 grid at a time, then quickly glance at the rows and columns running through it. Does your scan tell you that a certain number can't appear in the top two rows or the left-hand column, for example?

#3. Rule out possibilities with cross-hatching. Focus on one number at a time and find its givens in the grid. Because each number can only appear once in a row or column, you can imagine (or very lightly draw) lines crossing out the empty cells in the row or column where it appears.

5	7		9			4	2	3
3	1	6			4	9		
9			7	5		8		1
1			2		8			
			1			6	5	4
4	9	7		6				2
				8	1	2		
	2	1	4	9				
			6				1	

Example of technique #3:
In the example above, the gray lines show the only possible location of the 1 in the center box on the lefthand side.

5	7		9	1		4	2	3
3	1	6			4	9		
9			7	5		8		1
1			2	341	8			
			1	37	3579	6	5	4
4	9	7	35	6	35			2
				8	1	2		
	2	1	4	9				
			6				1	

Example of technique #4:
By writing out all the possibilities in the center box's cells, you see exactly where you must fill the 4 and 9. After that, you know where 7 must go, leaving only two cells unfilled.

#4. Write it out. Whether you're just starting out or working on a challenging sudoku puzzle, it can be helpful to jot down possible solutions—referred to as notes—in the available cells. To keep yourself organized, write the numbers in a 3-by-3 grid within each cell. Use a pencil so you can erase notes as you eliminate possibilities in the grid.

#5. Search your notes for single, obvious possibilities. If in your notes you find only one of a given number in a grid, row, or column, remember that the answer has to be that number—even if there are other notations present.

#6. Look for doubles in your notes. Next, look for obvious pairs in any two cells, also known as *naked pairs* (because they're not covered up by any other notes). If two of the same notes exist in two cells in the same row, column, or grid, then the answer to those cells must be one of those two numbers. That means that if those numbers appear in any other notes in that same row, column, or grid, they can be eliminated. In any cell where you only have one note left, you have your answer.

Example of technique #2: *The 1 in the first row of this puzzle is the only possible option, given that the third row and the third column already contain 1s.*

8	1	7	3			5		9
6	3			9			8	7
	9		8				1	
4				5				1
	7			8			6	
	6			4			5	8
3	5				8			2
7				3				
9	4	1			2	8		5

Example of technique #5: *For the second cell in the bottom row, 4 was the only possible number. The cell two rows above it could have only contained a 4 or 5; with the 4 already added below, 5 becomes the only possible value.*

#7. Find pairs *hidden* in your notes. In the previous example, the pairs in your notes were obvious—or naked. In this example, the pairs are there—but hidden by other notes. Don't let the other notes distract you: If a pair of notes appears in a grid, column, or cell, it means that those numbers *must* appear in those two cells. If the cells include other notes, you can remove them. Voilà! You uncovered a *hidden* pair.

#8. Look for threes. Just as you spotted pairs using techniques #6 and #7, you can also look for three repeated notes. For example, if there are three cells in a grid, row, or column that repeat the same three notes, the answers for those cells will contain one of those numbers. You can eliminate other notes.

#9. Find Xs. In this case, X marks possibilities. You can ignore the 3-by-3 grids with this advanced strategy! Known as the X-wing technique, this involves searching your notes in rows and columns. Look for an instance in which the same note appears in parallel rows and columns, as the 4s do in the example on page 14. If this occurs, you know that the final placement for those notes will be on the diagonal, because numbers can't repeat within a row or a column. You can then use information from the rest of the puzzle to eliminate possibilities and home in on the correct answer.

#10. Go fish! Swordfish, that is. You can ignore the 3-by-3 grids with this advanced strategy! The swordfish technique can be used only when a certain pattern can be found in the grid—and not every Sudoku grid has this pattern. For this technique, a single note has to appear in two or three cells in three rows or columns. As shown with the 8s in the example on page 14, if that pattern occurs, then the note can be eliminated from every other instance that's not in that row or column.

Example of technique #9: *In this partially completed puzzle, it is clear that the 4s will be placed in either diagonal. See the "X" here with the diagonal 4s. As a result, you can cross off the other 4s in the same columns.*

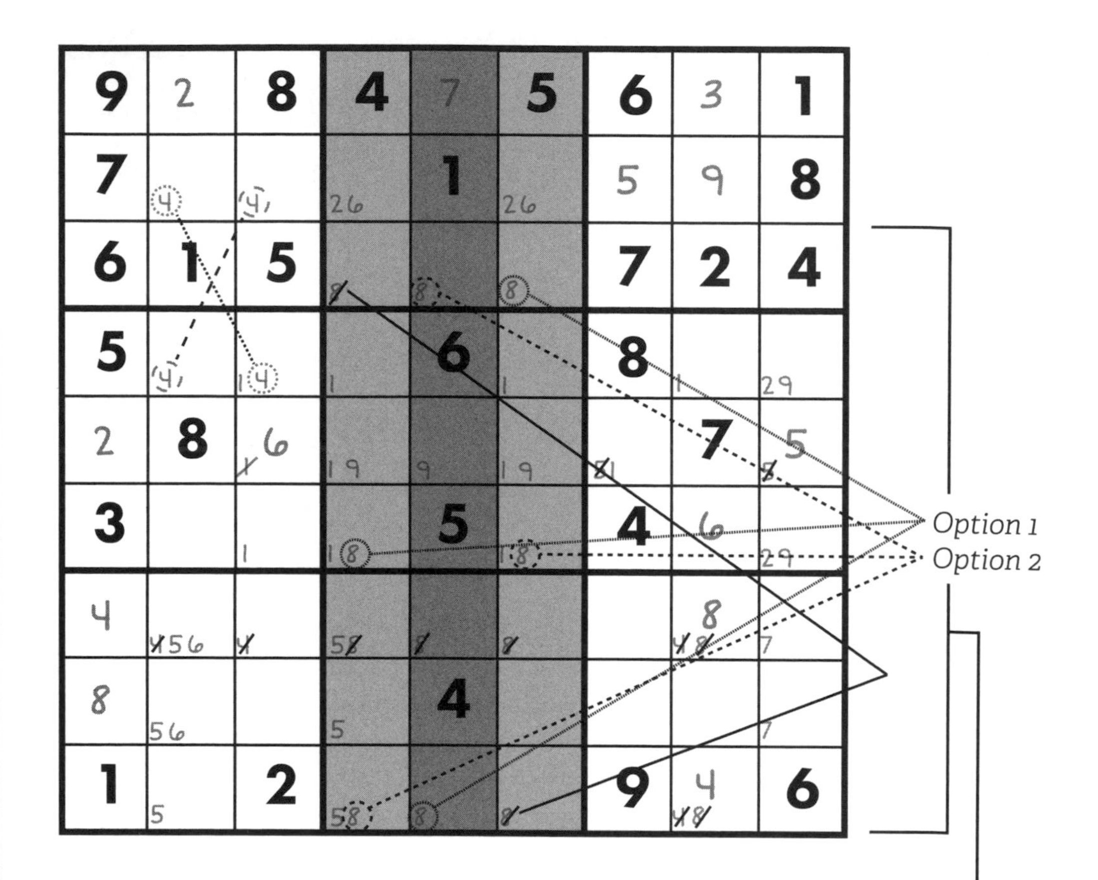

Example of technique #10: *In this partially completed puzzle, the "fish digit" is 8. As shown with the dotted lines, the 8 appears in three columns, so there are only two options to place the 8 in each row. All other appearances of 8 in notes in those rows and columns can be crossed out.*

GETTING UNSTUCK!

Sometimes, no matter how long you puzzle through or how hard you've been working out, you just hit a wall. It happens! When a sudoku grid becomes more frustrating than challenging, try some of these suggestions for getting unstuck:

- **Change up your technique.** Work your way through the different options above—try a new strategy! You can also search online for other tips and tricks.
- **Take a break.** If you find yourself getting frustrated, stop! It's OK to walk away for a little bit so you can come back to a puzzle feeling refreshed. You just might see something you didn't before.
- **Ask a friend.** Text a picture of your grid to a friend or family member who also plays sudoku. Maybe they can offer a new perspective.
- **Start clean.** Sometimes, you might need a fresh start. Erase your work, stand up and stretch, take a deep breath, and start again from square one.

HOW TO USE THIS SUDOKU CHALLENGE

The 90-Day Sudoku Challenge presents you with two sudoku grids to complete each day. The grids are organized into three 30-day sections that progress in difficulty from easy to medium to hard (plus one bonus chapter that has a mixture of all three levels!). As you work through each 30-day section, you'll stretch and strengthen your
sudoku-solving muscles—just like how you might increase the reps and intensity of a physical workout to build muscle.

While the book is structured to present a gradual workout over 90 days, the most important part of the challenge is to do sudoku *every day*. Even if you complete just one puzzle daily—from any section of the book—you'll reap the benefits.

So, approach this book in the way that suits you best:

- You can work on it straight through from day one to day 90 (and beyond) in the order the puzzles are presented.
- You can structure your challenge to increase in intensity over the course of a week: Ease into the

week with puzzles from Chapter 1 on Monday and Tuesday; move on to Chapter 2's medium puzzles on Wednesday, Thursday, and Friday; and then practice with the hard puzzles on Saturday and Sunday, when you might have more leisure time to complete them.

- You can skip around depending on what suits you on a given day: Need a quick workout? Do an easy sudoku in Chapter 1. In the mood to test your endurance? Pick a pair of hard puzzles in Chapter 3.

Now you have everything you need to begin the 90-day Sudoku Challenge. Ready? Take a deep breath, clear your mind, grab a pencil, and turn the page.

CHAPTER 1

30 Days of Easy Sudoku

It's as easy as 1, 2, 3 . . .

Congratulations on beginning the 90-Day Sudoku Challenge! For the next 30 days, you'll work through 60 "easy" sudoku puzzles. Each puzzle contains 35–40 givens. In some cases, the correct answer may just leap out at you—and that's OK! These easy puzzles are still doing important work; consider them the "light weights" of your workout. The goal in this section is to start forming your daily sudoku practice, begin building those puzzle-solving muscles, lay the groundwork for the bigger challenges ahead, and have fun while doing it!

DAY

1

5							8	
3	4		7				1	
7			1		2	4	3	9
9	5	3			6	1		
2	1	6				5		8
8	7						9	6
6		9			4	8		
	8		9		5	2	6	
4	2	5			8	9		

	6	9		5			2	
		4		1		9	5	7
3		5						
	9	6	5		4	2	1	3
5	8				1	4		9
	3		9	7	2	8	6	5
2	1							
			1		8		9	
9			2		5			6

DAY 2

		6	5		9	3		4
5		1		4	6		8	
	9					1	5	
				8	2	6		
1				6	3			
	3	2	7		5	8	4	1
2		9		5	8	4	1	
	1		9		4			
8		4	2		1			7

6	7	9			5		1	2
4	2			9	6	7		3
8	3			2			9	
			5	6			4	
5		4	9			8		
		8		4	2	1		5
9		7	2		3	6		
2			6		4			1
1		6				2		

DAY

3

					2	5		
1	7				6			2
			5	3				1
8	6		3	4		2		
3								
	9	7	2			8	5	3
	3		4	9	7	1		5
5			6		3	7		9
	2		1			3		

		2	1			9		
					6	2		7
9			8	2		5	4	3
				1	4			
8	4				2			5
		3	7				2	4
3		8	6	7			5	
	1	4	2	5	3	7		8
5		7		9				

		9	1		5	6		8
				3	8		2	
7	8	6	4	9		3		
8		3	7	5	9			
5		1	3	8	6	4	9	7
	9							
9	7			2			1	
	6	2			4			
1	5						3	4

	8	2			5	4		6
		6					9	
	7		9					
		8			3	7	6	
	1			8	7		4	
	6		4	2	9	5		
6			8		1			5
	5	9	7	3		1	2	
3		1		9	2		8	7

DAY

2	4			8	9	6	1	
		8		3	5	4		
		6	2		4		7	
			4	9		2		8
7			3			9		
	9	4	8				6	
5	2	3		6				4
		7		4				
4		9					8	6

6	8	7	2		1	4		9
	3			9			6	
4	9	5	3	6				2
		8		7	2	6		
	2		5	1		8	4	
		4						7
8	4	9			7	3	1	
						7	2	
		2	1		3	9		

2			9				4	
9	4		5			1		
			4	7			2	
4	5	3						
		6	8	3	9	4	5	
		8	6				7	
3		9	2		8	5	6	4
		4	7			2	3	8
				5		9		7

	4	9	1		6	8		3
				4				7
7		3	8	9	5			
		7	4	8		1	6	2
6		1	5		2	7		4
			7			5		9
		5			8			1
8	7			1		3	2	5
1				5				

DAY
6

DAY

7

9							4	
6	2		9	4	3			
		7		1				5
2	7	6				1	5	
4	5		1		8	7		9
	8			5		4	6	3
3	6	2			1	5		4
	9	4			5			1
5				7			3	

5	4		1		7			6
6	7			3	5			1
						8	5	
4					1	6	9	5
	6	9	5				8	4
	1	5		9				
		7	2	5		4		
2		4		6	9		1	
9			3		4			8

4	9	1	6			8		
6	8			4	5	1		
	7		1		8	6		9
	3		2	8				
		9	5		4			1
2	1	5	7			3		4
3	2			6		4		
		6	4	1		2		
1	4	8					6	

6					9	2		
	7	8		2		6		
3	2				7	4	1	
5		2			8			7
				5	4		6	9
	8		7	3				
4			6	9			8	1
				7			2	
	9		8	4	5		3	

DAY

9

			4	8	6	9	2	
	2	8	9	5	3	6	1	
			7		1		5	8
1	8							2
9		3	1	7	2	8	6	
	6		8	9				4
						3		
	9		5		8	2	4	
		4	6			5		

4		7		3	5	8		2
	1	6				9		
							1	
	3		2	4		5	9	6
	5			7			8	1
6		8			1	7		
			6	8	3	1		9
5			4		7			8
1	8	3	5		2	6	4	

8				4		7		
	1		7	2	6		9	4
			9	8	1			2
2	7	1	8	3	4		6	
				7	2		8	
4	6	8			9			3
	5	3		6			2	
		4			3			
9		6	1	5	7		4	

3	6	9	8			2	1	5
						6	7	3
2	8	1	3	5				
5							3	
9	4		2	8	7			
		4			8			6
8		5	7			1	9	4
		2	9	4		7		

DAY

11

			6	3	9	1	7	
	4		8	7		9		
		6	5				3	2
	9	3	7					
1			4	5	8	3		
4		8			6		5	
	1					6		8
	3	4	9		5			
2	8				7			3

		1				4	5	
4			9	3	5		1	7
7		8		4		2	9	
1			8	6		3	4	
	8	4						
5				7		9		
8		5						4
		6	7		9			2
2					8		3	9

DAY

12

	7	4			8		9	3
5		6		9				8
	9	8		1			7	5
	5	2						1
	1					3	2	7
	6	7		2		9	5	4
9		5	6	7	4			
		3	1	8	9			
				3		7	8	9

	9		8	5			2	3
				2		6		9
				9	7		8	
6	1	7				5	4	2
8		9					6	
	2		6		5		9	1
2	8	5					3	
		6						8
9	4	1	2	3	8	7		6

DAY

13

		8	9					
			3			5	2	
6		7		2				
4						2	7	
3	2		5	1	7			4
	6	1	8		2	9	5	
	7				3	4		5
	9		1					
1	4	2	7	8		3	9	6

	1	6	5			8		
			3			5		9
	5		6	7	4	2	1	3
	8			3	6	1		
	3			5		6		8
	2	9	8	4				7
	6		1		7		3	
		2		6				1
1		3	4					

DAY

14

2	6				4		9	
		9	6		2		8	
7						6		5
	9		5	4	7	1	3	8
		7	9	8				
5			1	2	3	9		
9	7				8		1	
	2	6	3	7			5	
	1			9				4

8	6	1			4			
	3		6	2	5	8		4
4		5	8		3			9
		3		8				1
			4		7			
							9	
3				4		9		
9				6	8	1	3	5
1	8		5	3	9	6		

DAY

15

		2	6	3			7	1
	3		4	8	9			
		8	2				9	
			7		4		5	3
				9	1			
4	7	9			6	2		8
		4	1	6		7	3	
7						1	8	9
2		1				5	4	6

6			2			3	9	5
1		9				2	6	7
	6		4	2	5	7		
		4		7	6	1	2	8
			8		1			4
7	5	3						2
	9	6		5	2	8	1	
	1				9	5		

DAY

16

		2		3			4	
7							2	
	9	3		4		5		
8				7	6	4	3	
4	7	6	1		3			2
9	3	5			4	6		
		9		6	7	2		1
6				9		3		4
	5	4	3	8		7		

		5	6				8	
6	9	3		8	2			5
	8	4			9			3
			8			7		
	3		9		7		6	8
5	7	8	4	6		1		9
	2		7				5	
8		6		1	5	2		
3	5			9		8		1

DAY

17

	1	4	6	5			3	2
2	5		3		8	6	7	1
3								4
6				8			9	
	4	1				7		8
9			2		1	5	4	6
	6	5		9				
			7	3				5
8			5					9

			5	1			9	
	8	5	6	3				
		4	9	2				7
			4	5			6	
6	1	9		8				
	4	2	7		9	3	8	1
4	3						2	5
	5	1			2			6
			8	7	5	4		3

9	4	6	1			5		8
8		2		5	4			3
5				6	8	4	2	
		9		7	6		4	
1				9	5	2		6
				2	1			
	9	8	6			1		
	2			1	7	9		
4				8	9	3	6	

3		2		1				
8		5	6	7	2		4	3
		1	3	4		5	2	
		4	7				6	8
	8				6			
	7	6	4	8	3			
6			2		4	8	5	
2		9		6	5	7		
	5					2		6

DAY

18

DAY

19

8	7						4	
	4					2	6	5
	2	6			9	8		
	1	8		2		6	7	9
2	5		3			1		
6					1		2	
	3	2	6	1	4	7		8
		9			2	3		
	6		9	3	8	4	5	

		2	3		8		7	1
7							2	3
	9			2		5	4	
2	1		4					5
	7	6	5	9	3	1		
					7	6		
1		8	6					
			1				6	8
			8		2	3	1	7

DAY 20

	4	9			6			1
7		2		8	5			
6	8	5			1			
				5	9	8		
		3	6	2				4
5	2		8	3	4	6	7	
			7			9		
	5			6	8	2		7
9		7					6	

	3	2	9					
	8				4			
9				2	5			7
	5	7	1		3	6	9	
	6		7	9		5	1	
	9	3		6				8
3	2	5	4		6	8	7	
	1							4
			2					5

DAY

21

	1	7			8	6	3	
3				1		4		
			9	3	5		7	8
7		8					5	
			5					
	3	5		7		9		2
	7	6		5			9	
9	5	1		6	2		4	
4		3		8		5		1

1			4	2	3	6	8	
	4				6	5		3
	2	6				1		
9				1		4	3	
	1	3		9	4	8		6
	7	4		3				2
	6		7			3		
4					2	7		1
			3	5	1			

6					1			9
				3		5	1	
			5			2		
3		6	7					
	4	8	6	5		3	2	1
	1	5	2	8	3			7
1	3	2					7	5
5	8			6				
	6	9	1			8	3	2

1	7						6	
5	9	8	4				1	7
		2	1	8	7	9	5	4
				7	6	1		9
6	8	9			1			
2		7		5			8	6
9		3				6		8
	5					2	4	
	4		7		3		9	

DAY

23

		6		5	3		9	
3	9	4				8	1	
				9		3		2
	7				5		2	4
					9		3	6
	6		2	1		5		9
	3	7	5			9	4	1
	5	9		3		2	7	8
	4	2	9				5	

	7			5		8	6	4
			8	2			3	
	8		4	1	6			
7	9		1	6	4		5	
2	4				3	6	1	
			5			4		
5		8	9					6
		6		4	5	1		7
4			6	3		5	9	2

		1		8	7	9	3	2
3	7		2			5		8
						4	7	6
		4		5			8	9
5		6		3		7		4
1			7		8			3
	1	3	9			8		7
2		7		1	6			5
		8		7	3			

	7	9						2
			9	8		3	6	
					4	9		
		5		7		1		
7		3	4	1	2	5		8
8		1	6				2	
	1		3	4				9
			7		1	2	3	4
		4		9		8		7

DAY

25

		5			3	7		2
		7	1		6			
		8		9	7		1	5
		9		8		3	7	
	8	3	7	2	9		5	
			3	6	1			9
7					2			8
4			9			2		
			4	7				6

			2				8	5
		7			3	2	9	
				4	1	6		
	7	5	3					
		9	4		5			
8		1			6	3	5	2
7	8	6	9				2	
9		2	6	5	7			
	3	4		2	8			6

3		6	7	5	2	8		4
7	5			4				9
	1					5	3	
8		4			5	1		6
		1						
			3	1	4	9	8	2
			9		1	3		
1		3	4	8				5
2	8	9		3	6			

	5					6		3
		3	4			1		5
6	8	1	5	3				
		8	7	5	3	4	9	
3	4		6		8		5	1
	9		2	1	5	8	3	
			8			9		
8	1	7	3		9		6	

DAY

27

	3	8		6			4	2
		4		1				
	5		4	9		8		
4	2		8		1	6		
5	8		9					
		9	6		5	2		
1	7	5	3	8	6	4		
			7		2	5		
		2						7

3		7	1		5	6		
1	4	2		9	3	5		
		6		8		3		2
7		5			4		3	9
4	9		5	7		1		
		1						7
	3			4	1			
5		4						1
2		9	7	5		8	4	

DAY

28

		1		9		3		4
3		9	4			2		
5	6				7		8	1
2	4	7	9		8	6		
	8		1	4			2	7
		5			3	4	9	
7	5				9		4	2
		8				7		
	3				4		5	9

				7	5			2
5	3		8	4		1		6
8	9		2	1				5
		5	4		8	7	2	
			7		1	6		8
7				2	3	4	5	
9			1	8				4
				6		8		3
6	8		9		2	5		

DAY

29

8				2		7		
					3	2	6	
5	2			6	9	4		8
9	8					6		7
1							5	4
	4	5		7	1		2	
2		4			6			3
3		8				9	4	6
7			9				8	

2			8	4			9	1
8	1	9	5	2				
	4			9			5	8
1	8		3		2	9	6	4
			4	5		1	2	
4							8	3
					5	6	3	
	2	1	7	3				
		8	2	6	4	7		

DAY

30

			3			7	8	
	3	2	5			9		
	7			1	2			3
7							4	
5	6		2				3	
	9	1	6		4	2		5
3		6	4	5				7
2			1	6		3	9	4
	4	9		2			6	

6		8	3	2			1	4
9					8	7	6	
	4	3			9	8		
		7				4	9	
3		6		5	4		2	7
		4		7				
	3	5		9			7	
7	2				1			
		1	5	8		2	3	

CHAPTER 2

30 Days of Medium Sudoku

Let's take it to the next level!

By completing the first chapter of this book, you've finished a third of this challenge. You're well on your way! For the next 30 days, we'll build on the foundation you've created. You'll puzzle your way through 60 "medium" sudoku. These games have 30–35 givens, and you'll notice fewer cells where the answer is obvious. If you feel stuck, take a deep breath and try not to get frustrated. Turn back to page 7 and review the solving techniques there—they just might come in handy.

DAY

31

		7	8	9	4	3		2
				5		7		
	4						6	
3			6	1			2	7
					5	8		6
4			7				5	
5		9		6		2		3
	6	3			1	5		
	2		5		9			1

			5	1				2
		1			2	5	4	
		2	8	3	4	6		
9		8						6
7			2	6	9	3	8	
					8			4
	2				5			3
	4		1		3			
8		5			6	2		

7	3			2	1			
		1					7	
5		4	7					9
						4		
1				5	9	7	6	
9			2	7				1
	9		1					7
4					7		8	3
8		6		9			5	

4	5		2	9	8	7		1
1		9				2		3
			3	1				
2	4	1	5	7	9		8	
8		6					5	
9			8		6	4	1	
			4		2			
			7				3	
7					5		2	

DAY

32

DAY

33

		4						9
6							1	
7		8		9		2		
1	4				8			5
			5	2		3	4	1
3	7		9					2
4				7	9		2	6
				3			9	
		6			1			

	1	8		7	9	5		3
3	7		8	2				
			3	5	4			
6	4			8	2			9
8		1			7			5
9			1	4				6
		6			5	4		7
	2					9	6	
	9		4					

	6				4	5		
1	4				8	7		
7		9	1	2		6		
	7				2			
4	1						3	
3		5		8	7	1	6	
6	3				9	4		
5	8				6		1	
2	9		8				5	

	2	9	4		7	8	3	
	3							
1							4	6
2			8		1	3		
		5	9					
	8			2		4	1	5
8			3	1		9		
7	9	3	5	4		1		
	6					2		

DAY

35

			9	2		6		4
	3			4	5			
9						3		
					4	9	8	
		7	6		1	4	3	
	1				9	2		6
	8			1			9	3
4		3	8			1		2
		9	3			7		

	1				5		7	6
	4					3		2
	2	7			1	8	4	5
3						7	6	
5		6	1	8			2	
				3	6	5		
	6		2		3			
2					4	6		7
			6			2	8	9

6	8				1			7
1	9	7	6	2			4	
		4	9		8		6	
9			3	1	4		7	5
					2	8		
					9	1		2
3	1	2		4				
				3	7			
8						4		3

4	7	8						2
5		9				6		
		1			5	7		3
1	9	2			7			8
8			6		1			
6		3		2		5		
9				8	3	1		6
		5	7	6	4			
		6		9				5

DAY

37

8				2				
5		3				7	2	9
						8	3	4
6		4	7					2
		8		1		5	4	6
1		5		8		3	9	
3	7					4		
			4		1	9		
	5	1			8	2		3

		8	1	4			9	
9		1		5	8			
7	4	5	6		3		1	
6		3	7	1	2	9	4	
		9			4		7	
3	7							5
	1			7	6		3	9
8	9			3				

DAY

38

		8						3
			3		8		7	9
	4	3		6	9		5	
4	1					5	3	
	3	6					1	
				8	3	6		
6		4			7	3		
			6	3		9	4	5
3	2	5		9				6

2				4		6	8	7
6	3		8		9	2		
				2				3
		3	9			5	6	1
8	5			3			4	
7			1	5			3	2
5					2			
	8	2	7				5	
		7	5				2	

DAY

39

2	6	3						7
				3	9		1	6
			2					
			8			9	6	4
		9						3
4			3		5	7	2	
8			6	7		5		
6	9		5		3	8		1
		5					7	2

		8	3					
	5						9	2
6	9		7	5			4	
	6		8					
	4			1			8	9
			9			6		
5				7	1		3	8
				9	5		7	4
4					3	1		

DAY 40

			6			2	1	8
3								
1							9	3
8		3	9	1		7		
4			7	2	8		5	1
				4				
	8			6	1			5
5				9		8	6	2
			5	8		1		9

			9		8			5
		9	2	4	7	3	6	
	2							
	1		6	2	4		8	
	3	2	1				5	
8								
2		1			6		3	7
4	7					6	1	
		3		7		8		2

DAY

41

	8	5	9			3		1
2							7	
				3		8	5	4
9	3		2	7			6	
			1			2	9	
5			4		9		3	
3	1	7		2		9		
		2			7	5	1	

3	6	1	5				7	
5	2		6	1			3	
					3	5	6	
	3	9		8			2	
				9		6		
	8	4			1			3
8	1					7		
7		6	1	5	8	3		
9		3						

DAY

42

6	5	4				7	3	
	9			2				1
				7			6	
	4		7	6			1	5
		1	5	4		9	8	
	3	6		8				
				1		4	9	
4		7	9		6		5	
		9		5			2	7

5		9		7	3	2	1	8
		1		2				9
4					1			
	1			3	2			7
	8			6			4	5
6	5		7			9		
3					7	8	9	6
				4		5	3	
		2					7	4

DAY

43

5	9				1	2	4	
	7				8		1	6
1					9	5		7
		3		6				5
						8	9	
	5		8				6	
				8	6	7	5	
8		4			7	3		
7				3	4		8	

	1	9	7	4			8	
8	4		1	2			3	
					6	5	1	
		5	4		7			
4		3		5	8	7		
2	8				1			
	5			7		4	9	
9		8					7	3
				1				8

DAY

44

		3		5	4		9	
					3			6
	7				8		4	3
			6		9			
			5	4	1			7
		4		8		1	6	
6		8	4			3	2	1
	2		8	3		6		
9			2			4	5	

5				9		1		
	2							
	9	7	8	5				
		3	7	2		4	8	1
		9	3	6		7		
	4		5	1		9		6
2				8	7			
9			1					5
		4		3		2		

DAY

45

				4				6
7				3		4		
			7			3		
1		4	2	7	5			3
8	2	5			1			4
6		3		8		5	1	
		6	9	1			3	7
9		7	8		3		5	
				5		9		

				3	4	2		
	4	5	8	9		3	1	7
		3		7		8		6
		2	6					
4				5			6	
3	6	7			1	9		
	9		4		8			
		4	9	1		6		
5				6	3			

DAY

46

5				6		2	3	
	1					5	7	
	6		5	2				9
	8	3	2				4	
4	2			7				3
	5	1				7		2
	7				9	3	8	4
		5			2	9	6	
	4		8				2	

							8	
6					5		1	
	9	5				2		6
2			8		9		3	
9		7		2		6		1
	3	4	1					8
4			9	7	8		5	
			6	3	2	8	9	
					1	3		7

DAY

47

			6		9			8
					1			4
7		1			3	2	9	
9	4					8	2	
					4	9	5	6
		5						3
		3		5	8	7		
6			1		7	4		9
8		9		6	2			1

3	4				5	8		1
9								
8		1		2		9		
	8	3	6				1	
2	1			7				5
4			3	8		6	2	7
					6			4
6		7	5		3		9	8
				9			6	

DAY

48

9	2	5	3	8	7		1	
						7	3	
		7	4	9			5	
	7							
	4	8			3	1	9	7
6			8	7	5	4	2	
				2		3		
	3			4		9	6	2
		2						5

		5	6	8		7		
			5		9		4	3
7		2						
1					5	3		9
	8		4					
5				9		4		
8				2	4	5		6
	7	9		5	6	1	2	8
				1				

DAY

49

7	2		8		1	6	5	
	3			4	6		7	
	5	6		7		4	1	
			1				9	
4		1		9	5			7
5		3					8	
2			3					
					4			5
9		5	6		8		3	

	5		4	7		8		
		2						1
		8		6			2	
1								3
	2			3	8			
6	9		2		1	7	8	
				2	3	9	7	
				1	4		3	
	3	5	7			1		

7	3					4	5	
1	5			9			6	
				7		2		
2			7		8			4
5							8	1
3		8			4		2	
	6		2			5		8
		5		3				
	1	7			9			2

		2		3			5	9
	5					8		2
7		1	2				6	
	6	4				2		
			9	2	6			
	3	7				6	9	8
	9	3		6		5		
4		6		8	1			
	2	5		9	3			6

DAY 50

DAY

51

		7		8	9			4
6					1		8	
	8		5			1		7
2	7			3	5		1	
4					2		3	
	1		9	4			7	2
	2		4			3		
					6			1
7	4						2	9

9		8		2		6		
			9				2	
	3		7		6			9
5	9	2			7			
7			4	8	9	2		
			5		2			7
	2	5			4	7		
	7	6				4		3
3			8				6	

DAY

52

		2		1	5	9		
							4	7
			2					
8		7				6	9	1
1	4		9	8			2	
	3		1				8	5
2		3	4					
6	8	4		3	9			
5			8	7	2			

	3	2				6	9	
8		5	6		2	3	4	
			1		7	8		
6				1				
7		3	2		9	5		
				6	4			
		8		7				3
5					6			8
3						4	7	5

DAY

53

6	5	9	3	2				4
			5	4			6	3
		3			6			
5	3							
	8			5	4			
		6	9		3	5		
			6			8	1	
					2	3	9	7
	2	1	7	9		4		

		9	6	8			3	4
		4	7	9			1	
2				3			9	7
				7		9		
	1	5	2		9		6	
	3	7		6		1	4	
		3		1				9
7				2				1
4	2					8		6

	4	7	5	8		9		3
9	2		4			1	5	
6		5	1			7		4
8		2	6		7	5		
5			9	4	1			
	7	8			5		9	
					9	4		
2					4		3	7

			4		5	1		6
		4			7	5	3	
5				6		7		
			8		6	9		1
	1		5		9		7	
	9			7	3	8	2	5
	4		7					
8	5		6					7
	6	2					5	9

DAY

55

		9	1					6
8		1				3		
	6		4		5			
			5		9	1	3	7
				7				8
				8		5	2	
	7			2			5	9
6			9			4		
			6	5		7		2

		5						2
7			2		4			
2	3					6		9
			1		6	7	9	
6				7		2		
			8			3		
8	5		7			9	2	3
	7			5	2			
3	1			6		5		

DAY 56

9	6	7			5		4	
5			9		2	8	1	6
		8	6					
4				9			5	1
	1	5	2				3	
3	7	9	5		1			4
	9	1		2			6	
2					3			
		3				4		

	3			6	5			
8			2	3	1	9		
				7		8		3
	4	2	7					
6	7		5			2		
			1	2	3	7		
					2		8	7
4	6	7	3		8		1	
2				1			4	9

DAY

57

7			5	9		2	6	
	4		8					7
2		5	7			4		1
9	7		2	6		8	3	
5	6							
		8					2	
		3		7		6	1	2
6		7					5	8
		9		2			7	

	2	7			6			1
1	3	9	2			5	6	7
		6		7			8	
						1		
	1		6				7	5
		2	1	9				
2					9			
9	6	5	3				4	
4	8	3		2				9

DAY

58

4					9		2	7
9							8	
	7			2				
	6		7				3	9
	9			3	5	7		
				9		2		
		1	3	7	8	6	9	
6		7	9		1		5	2
				6	2	1		

		3	4	7	5	9		2
6	7	2			1		4	3
5				6			1	8
	3			8			2	5
	2	8			7			
9		5					7	
2				1				9
		6	7		8			
			9				3	

DAY

59

			4		6		5	
		5			8		9	
	1		9	7		3	8	
					7	6		
9				2	4			5
7					1	9	2	
2		1		8				4
5			7	4	2	8	6	
6			1	5		2		

	9							
4	8	6			9	7		
3	1	2		7		6	9	
	7	8				5		
1		9	3	5			8	
				8	4	1		
5			6		7			2
		7				3		
8			5		2			

DAY 60

4			2	6			9	
				5	3		2	
6	2			8		1	5	
5		1		2	4	3		9
		2						8
	9			3	7	2	4	
			6	7	8			
				1	5	4		
			3				1	6

		8	6		7			
	6	1	8	9				
7						8	9	
				7			2	3
			2		4			
5		4		8	6			7
	7		1	6				
6				4	3	1		
	4	9			2	3	6	

CHAPTER 3

30 Days of Hard Sudoku

Don't be intimidated! You can do it!

This is it: You've made it to the final phase of the challenge—and you're ready for it! You've powered through two months of daily sudoku, and in doing so, you've strengthened your problem-solving skills *and* your self-care routine. You're ready for 60 "hard" sudoku. In these puzzles, there are just 25–30 givens. There's nothing obvious here: you're going to have to muscle through to find the solution. Remember that you can find techniques on page 7 to help you figure out the answers. You've got this!

DAY

61

			3			6		
		4	9					
7	3		2	6	1		8	5
5							1	
	8		7	1	5		3	
	7							2
9	2		5	7	6		4	8
					4	5		
		5			9			

				8				
	7		5	3		1		
2		6	1		7			4
					8	7	9	2
		5				8		
3	9	8	7					
1			8		9	5		3
		7		1	5		6	
				6				

DAY

62

				2			4	
		5	8	9				
		1			3	8		5
	5	3			9	7		8
		2				4		
8		6	3			2	9	
7		9	4			3		
				7	2	5		
	1			3				

			8			7		6
			5	6	9	1		
				3		4		5
5		2	1		6			3
3			9		4	8		1
4		3		1				
		5	4	9	8			
6		8			5			

DAY

63

					2	8	6	1
		6	3	7		5		
					1			7
9				4			8	
	4		2		8		5	
	5			3				4
7			1					
		4		9	7	3		
5	9	1	8					

	8						6	
		6	1	9	7	8		
	1					5		2
	3		9			4		
		5	2	1	6	7		
		7			4		1	
1		4					5	
		8	3	6	1	2		
	6						9	

DAY

64

5				1			8	3
3	1		7	6				
		9		5				2
		4	6					
7			2		1			4
					5	8		
6				3		2		
				2	6		4	9
2	9			7				8

	1		2	5	8	7		
	2				3			1
8							2	3
2					4			
3	6						7	4
			6					8
6	8							7
9			1				5	
		7	8	9	6		3	

DAY

65

9			8				4	
5		2			9	7		
	7						5	
		3		8	6	9		7
			9		3			
1		9	2	4		3		
	1						9	
		7	3			6		1
	9				1			2

				5		1		8
			1				7	
	7	1			8		9	
6	8		9			4	2	
		4				3		
	1	3			4		5	7
	5		2			8	1	
	3				6			
2		9		8				

	7		6	4		2	3	
3		4			2			1
				3		4	6	
7		1			9			
			4			7		8
	9	6		7				
2			3			9		5
	3	5		1	4		7	

	8	7					5	
2		4		5				
1		5	9	7			8	
		9	7			8		
			8	2	1			
5					3			7
	5			8	9	4		1
				1		3		6
	4					5	2	

DAY

67

3							1	
			5	4			9	6
					7	2		
	2	5	9	6		8		
6		8				1		9
		9		3	8	6	2	
		7	8	1				
4	8			5	9			
	6							8

								7
			4	1		5		
6			7	9	3	4	2	
	3	4	9					
7	1			5			8	9
					7	1	3	
	9	3	1	4	6			5
		5		7	2			
1								

DAY

68

	9					8		
6					4	2	3	
4					8	1		
8			4	5	9			
	4	9		2		5	1	
			8	1	3			6
		7	9					2
	2	4	1					3
		6					8	

2		7		5		8		6
				9				
5			6		2			7
	2						4	
		5	2	4	3	1		
	1						2	
1				3				4
				6				
9		6	4		8	5		1

DAY

69

6					5		8	9
			6	1	7			
	1	5		2	9	7		
		2				6		
	6						7	
5		9				2		8
		4	7	5		8	6	
			9	8	4			
7	8							5

	3		6	4	8			
7	2	5						4
			7					
	9		1	2	7			
3		7		8		9		2
			9	6	3		4	
					6			
6						7	9	1
			2	7	1		3	

	7		6		3	1		4
	2					8	9	
			5					
7	1		8	6	5	4		
		5	3	1	4		6	8
					8		4	
	6	2					8	
4		1	9		6			

5			3		6	1		8
7		2		1		5		9
9		5			2	8	6	7
6	3	1	7			2		5
3		4		5		9		6
2		8	6		1			3

DAY

71

	9	8	7		3		4	
4			1		8			2
	7					1		8
			6		4	3		
				8				
		4	3		1			
6		2					1	
8			2		9			3
	1		5		6	2	8	

	3			6			2	
5								7
	9		3	7	5	1		
	7		5		1		4	
2								1
	1		6		7		3	
		9	2	1	4		5	
3								2
	2			3			1	

DAY 72

	4			1		6	5	
			5		3		8	
1						7		4
		7	4		9			
	8		7		6		2	
			1		5	4		
3		2						6
	7		6		2			
	5	9		7			4	

			9					
	7		1	4		9		
		9	8			3	5	7
	2				8			5
		4	5		7	8		
8			2				7	
4	9	2			1	5		
		8		5	6		9	
					9			

DAY

73

		8		1	7		2	
	5	7	6					
6					5	3		
7		5	1				9	
			3		9			
	9				2	4		7
		6	8					2
					6	8	4	
	8		2	3		7		

	3	5	2			4	9	
		9		6			2	8
			1			3		6
			7	4				2
5				2	6			
8		2			1			
3	4			5		2		
	6	7			2	8	1	

DAY

74

	8	3	2			1		
				1				3
1		9			4		2	
				8		4	7	6
	4						8	
6	3	8		9				
	9		3			8		1
3				4				
		7			5	6	3	

8	6		9	7	5			
	5					7		
		7		3	6			5
9			7				6	
	3						1	
	7				3			8
6			5	8		2		
		1					5	
			3	1	7		8	9

DAY

75

	3			4	9			
	6	7	1		2		8	
						7	1	
	2			3		8		
6	1						4	3
		3		1			7	
	5	6						
	9		5		3	1	6	
			2	9			3	

			6	1		7	9	
1					2	6		
		6					4	
	6	8	2				3	
7		4				8		6
	3				6	4	5	
	9					5		
		7	3					4
	4	1		2	5			

	7					3		
3	2		6		9			8
6		9				5		
			9	6				1
4			8		1			9
2				7	5			
		3				9		5
1			5		7		8	3
		5					1	

			9	4		6	2	8
					6		3	
	4			2	3		7	
	5			9	8			6
7			2	6			5	
	7		1	8			6	
	6		7					
3	1	2		5	9			

DAY

77

						5	9	2
	5				3		6	
2	1		5			3		
				1		2		6
5			6	9	4			8
1		7		8				
		1			7		2	5
	4		9				1	
9	3	5						

	3	9		2	1	8		
1								5
8		6		5		2		
4		5		6	2			
			8	7		5		6
		4		3		6		8
3								4
		1	4	8		9	2	

6		3	7			9		8
1					3	5		7
				6	9			
		5	6		4			3
		1				4		
4				5	8	2		
			1	9				
		4	8					5
3		6			5	7		9

		7		9			3	
1			5		6			4
		4			3			5
		3	8					
7	8		3		2		5	6
					1	3		
8			6			4		
3			4		5			9
	9			3		5		

DAY

79

2			1				8	7
3			7				1	
	1	7	8			2		
			9		1		7	
9		3			4			
			5				9	
		5			6	9	2	
	6				8			5
7	9				5			6

2		6						9
	3				7		8	
7	8	9	5					
	7			5		2		3
			3		9			
3		8		1			5	
					5	8	4	7
	9		4				1	
8						3		5

DAY

80

	4			1			5	3
			5		3			1
1				6			9	4
	6	7	4					
		1		3		9		
					5	4	6	
3	1			9				6
8			6		2			
6	5			7			4	

				6				
	8			2		9		4
7		3		1			6	
		5				6		
9	6		1		5		8	3
		1		3		4		
		7				5		
	1		8	5	4		2	
		8				1		

DAY

81

7	6			4				
		5						
	8	3			6			
		6			7	5		
8	5		2		1		4	3
		1	5			7		
			6			8	1	
				3	4	2	7	

	1			8	5	9		
			4					5
		7		6			4	1
		1			8	4		2
			9		7	8		
3		9	6					
4	3			7		5		
1					3			
		6	1	5			7	

DAY

82

5					6	9		
			1				3	5
	6					8	7	
		2		6	4		5	9
		3				2		
9	7		8	2		3		
	9	5					2	
7	4				3			
		8	6					7

		6	8	3	9	5		
	5	3		1				
9		4	5					
7		9		5				
	6						5	
				4		2		6
					3	4		5
				6		1	2	
		7	1	2	5	6		

DAY

83

	2						7	
9		1		6		4		8
	5		4		3		1	
6					9		3	2
7	8		6					4
5		2	1		8	3		9
				3				
3		9				6		1

			5		1		4	
		3	4		8	9		6
8		4						
2				5	4		6	
	5						8	
	3		2	8				4
			7		5	6		8
3		5	8		2	7		
	7							

DAY

84

3			6	9		5	1	7
			2					
					1	2	9	3
	3		8	1				
	1						8	
				7	6		5	
9	4	3	7					
					4			
2	8	1		6	3			5

5	1	3	9	6		8		7
					8			
9					7	6		5
4			5		3			1
				8				
1								4
8		4	6					2
			3					
3		1		7	2	4	5	6

DAY

85

	3			5			1	7
			7		3	5	2	9
9				1				
	7	6			2			
		9		8		7		
			1			8	9	
				3				5
4	9	5	6		8			
3	6			9			8	

5	2	1		4	3			
		9						
	4	7		2	5			
		8			1	3		
1	3		7		8		2	9
		6	3			7		
			1	7		2	3	
						8		
			5	8		9	7	1

DAY

86

	5	9		2	8			4
8						3		
2		6		7				5
6		8		3	1			
			2	9		5		8
4				1		6		7
		5						1
7			4	8		9	5	

	9					2		
4	8		2		6			
2						4	6	
		8		2	9			6
7			6		8			2
9			3	4		7		
	1	3						4
		9	4		7		1	5
		4					2	

DAY

87

1	3		8	9	6			
	8					1		
		4		5	7			3
4			3			2		
	7						1	
	1				4		5	
2			7	8		4		
		3					2	
			2	6	9		3	1

					5	3		8
		8		6		2	1	
	6	2			4		7	9
				3	1	6		
		3	4	5				
6	2		9			7	8	
	3	1		7		4		
7		5	6					

DAY

88

	4			1			5	
7			5					1
		5				7		
	6		4	2	9		1	
	8		7		6		2	
	2		1	8	5		6	
	1					5		
		4			2			9
6				7			4	

5	2	8			6			
			4			3		
						6	8	
3		1		4				8
9	6		8	3	1		5	4
2				6		9		3
	3	5						
		2			9			
			7			1	4	2

DAY

89

			1	9		4	5	7
					7		2	
	7			4	2		1	
	4			8	1			2
1			9	2			4	
	9		2	1			3	4
	3		4					
4	5	1		3	9			

5	8		7			6		
				3				5
6		7			5		2	
				6		3	4	2
	4						5	
8	2	6		4				
	7	5	9			2		4
9				2				
		4			7		6	8

DAY 90

9			2		4			
4		5				9		1
	8						2	
6		4	7	9				3
			4		5			
2				3	6	7		4
	1						4	
3		7				1		2
			8		1			9

					5	7	4	2
5			7	4		8		
					6			1
	6			7				3
	5		3		1		6	
9				5			1	
4			1					
		9		3	2			4
3	2	5	4					

CHAPTER 4

Bonus Sudoku!

Congratulations!

In the past 90 days, you've solved 180 individual puzzles, sharpened your powers of logic and concentration, and built up your working memory and decision-making prowess. You're a sudoku challenge champion!

Still, completing the challenge wasn't just about winning. You set out to dedicate time for yourself and your brain health every day—now, keep up your daily sudoku practice with the puzzles in this bonus collection. Here, you'll find a mix of easy, medium, and hard sudoku. Enjoy!

DAY

91

	4		7			1	8	
1	2		4	8				6
		7				9		
			1	5	4	6		2
	5					7		1
9		4					5	
2	7	1			5		3	
3	8			1	2			7
	9		3			2		

	6						5	
				4	9			
4	2	3			5	7		9
3			2	8	1	5		
2	5							6
8		4	6			1	7	
5	1	9	3			2		8
				2	4			
	4			1	8	3		

DAY

92

		7	5				2	
1			9			7	3	8
2				3		6	9	5
5	1			7				3
				4	3			7
3		2				1	4	
	2	5	6	9	7	3		
	9			8	2	5		
	6	3	4				8	2

7	6	3						
	4	8				9	6	1
9	1		2					7
	7	2				4		
	5	4	8	7		2	1	
1		9			4	7		
	2		4	5			7	
4						6	3	
5			9	6	3		4	

DAY

93

3		6		7	2		5	9
7	1	2	5			6	8	
					1		2	
		1			9			6
						4	3	2
		3						8
	9	8		3	6			1
	3	4			7	8	6	
	6	7		4		2	9	3

		9		1	6			3
					8	9	5	4
5		2						
	1	3		9			6	8
		7			2	3		1
				3	5		4	7
7	9		4	5				
6	8			7	1	4	3	9
		4		8	9			

		1	5		7		4	
	7		8	6		3		
5			2		1	9	7	8
					5	8		4
		8	4	1	2	7	6	
7						1	5	
	3		7	9			8	
8	1				3	5	9	7
	9		1	5		4		

	3				7		1	5
	7	1			5	2		4
4	2		1	9				7
7		2	6					3
5		4		1	9			8
3	1	8				5		
	8	3			4			1
9					1			2
1	4			2		8		

DAY

95

8							1	
3		1			5	7	6	
7	5		1	2		9		4
		8	6			1		7
2	1		8	7	4			
6					2	8		
9	2		5		1	4		
1	6		4	3			5	9
	8		2					

		8			4			
		2	9		7			
6	1	5		3			9	7
	3	7		9		6	2	
		9					7	
8	2		7			5		9
	5	6	4		3			1
			1	2			4	6
1	7	4	8	6	9	2	5	

DAY

96

6		9	4	5			1	3
7		3	1	6	8			
2				3				6
		2	9	8		4	6	
4				7		3	8	
8			5			1	9	
	1	7				2	5	
	6	8				9	3	
		4		9	5	6		1

	7	1		5				3
				7	1			9
9			4	8	3	1		
		4	1	6			2	
	8	3	5		7		9	
1	9		8	3	4			5
			7		5			
		7	2	1	8	9	3	4
	1						7	2

DAY

97

3	9			1		5	8	
		5	2			7		3
	7	1	4	3	5		6	2
	5		9			6	2	
	4		1	6	7	8		
7		6		8		4		1
9		7	8					
	8							
	1		3	7	9			

	4	7	5		1			6
6		1	2	7				4
8	5							1
	9				2	3		8
	7	8			3	9		
4		3	1		8			
	6		3					
3	8		9			4		7
		9				5		3

7				5	6		3	
6	3			9			5	
			7					2
	6	5		4		8		
8	4	7		6				9
9	1	3		2		5		6
4		6	9		3		8	
3		2	4	8	5	1		
1		8	6	7				

3	2			5	8		7	1
		7		2	6	5		
	4		7	9	1		3	
6	1		2	7	9	3	4	
		3						
8		4	1	6			5	7
	3			1			8	9
9			8	3				
7		8			2		6	

DAY

99

9	7				8			
2	4	5				6		7
1	3		7	6		5		
7			4	8	5	2		
4			6	1			7	5
	8	9		2	7			4
8		4	9			7		
	9			7			5	
6				3	4			

8							1	2
			1	8		5		9
	5	1					4	
	2	6		5	3			1
3		7	8		6	9	5	4
4			9		1	6		
6			3			2	8	
1			2	9	5		3	6
5				4		1		7

DAY

100

8		3	1		5			
			4	8	7	1		
1	7		9	3	2			
		7		1	6	2		5
	2		5		4	9	3	
5	1		2	9	3	8		7
	8	6		4	1		2	9
4		2						
					8			

	8		6			1		3
				4			8	9
9		5		8	7			
	9	8		1	6	5	4	2
					3			
	4			2	8		7	6
	1		2	7	4		3	
6			8		9			
		3	5	6			2	

DAY

101

	1		7					3
6				8	3		1	
	3				6		9	5
			4				6	1
	7						4	
3	4		8	2				
		9	3	6				
	5	4		7	8			6
1	6			9	2			7

			6		2	3	9	
	8		9	4	7	6	5	
	5					7		
		5	1	8			7	9
		3	2			4		
7		8					3	
2	9		5	1				
		1			6	5	2	7
			7			9		

DAY

102

1	6							
		3	8		2	4		
			1		9		3	
4		7	3	8			5	2
		2	9			3	6	4
	9			4				8
	2			9		7	4	
6		4	7	5				1
			4				8	3

						4	2	
	9			2	8			
	2		4		9		5	
1	6			5			4	
	4		1	9	3			
5			6	8			1	
		5	8			6		
6		3					9	
7	3	6	9	1		2		

DAY

103

9		4			1			
		1	7			6	2	
						9		
	7	8		9				
1		5		6		3		
	4	2			8	1		
4				7		2		9
	2	6				7		4
		9	2			5	6	8

3	4	8			2			
	5	6	4		1	8		
		9						
	9		3		4			
1		4	2				3	
			5	9	6		4	
6	8				9		5	3
			6		5	2		
4	7	5	8			9		

					4	7	1	
	4						9	6
	2	8			9	5		
8	6		9	1	7			
		5			2		6	
9	7		4	5				
		4		9	3		7	
	8		5			2		
3		1		2			5	4

5	1	8		3			2	
			6		8			3
			4		2			7
		6		2		7	9	8
	5		8		4		3	
		1	9					
							1	
	8	9		4			7	2
		2	1		5			4

DAY

104

DAY 105

	9	5				7		
		6	8	1		9	4	3
		3	9		7	5		6
2				8	3			
	7	4						
6				9	4	3	5	
	3		1	7		4		8
9		7				2		
		8	2		9			

	7		5					6
	9		6	4				
	6	2				4	9	
	2		3	1	8	9		
					5		2	1
		3		2	4			8
8		7	1					
				8		6		3
9					3		1	

DAY

106

			9	6		2		
	1		5	7		3	4	9
			4	3			1	
6	3			8				
		9	2			7	5	
		5			9	8		1
4		2					8	7
				2	7		6	4
			8		4	5	2	

	3					6	9	
9	5	8	6	7		3		2
		6			8	1	5	7
		4		1	6	5		
			5	8			7	6
6	7				3			4
							3	
7			1	3				
	2			6				1

DAY

107

5			1					
6	8		5		9	3		7
9				6	4			
		7		4	2			
1	2	6		5	3	7		
4			7		1	6		
		9		3			6	2
3				1	6		7	9
					7			8

	9			6		1	7	3
6	5		2		3	9		
4				8		2		
		4	7		2		9	
	3		8				5	
			4	3	6			
				2				4
	6	9					2	7
				1	7		3	9

9		3						
	7							
5	1		3			4	2	
8		7		6		1		2
		2	8	7				3
		1	2		5	8	7	
3		5		2				
1			6			3	9	
7							5	

	9				7			
	3	6						7
	5		1	6	9		3	4
3	4		9	7				
	8	9		1		6		2
				8				9
	2			3				
							1	8
6				9	8	5	2	

DAY

109

7			8	4	9			
4	8	5						1
6				5				
2			3		8	6		
	1	9			2	3	7	
3		8	7	1	5			2
		6						9
						4	2	
9				8				

	2		5					1
		5				2		
			2	1	4	3	7	
5		1				9	3	
		9	1	4				
	8	2	9		5	1	6	
	4	3			1	7		8
		7		8	9	5		
	5		3	2				

			4			8	6	
					6		5	4
					1		7	
5		4	8	1				2
1	2	7	6	9				
8	6				5		1	7
2	7			5		9	8	
4	5							6
		1	7		8			

2	8			3	9		6	
7							5	8
5			8		7			2
		8	6		2		1	5
6			4			9		
1	2							
		2				8		
		6				5	3	
4		5	3		1			6

DAY

111

6		3	4			1		
	7				2			
9	5				1	6	8	
			2	6	9			8
			5	7				1
		2		8		9		7
2			8		6	7		5
	6			2			9	
		5	9	4			1	6

				2	5	6		9
						4	2	
		4	7	9	8		3	
7			6	8	1			4
		3	5		2	7	6	
8							5	
6			1			2		
				7	3	8	1	
	3						7	

	5			2		6		7
8			9		7			3
9	7	2	6	3	4			
1	4	8						
					1	7		5
5	6			8	3		1	
6		3	8					
					6	2	3	8
	8		3		2			

			6	7	2	3	4	
4								
	6	2		5	1			
3			7		6			
2	4	3				8	7	
	9	6		8	3	4		
1		7					3	
6	3	1		2			5	
				6	9			

DAY

112

DAY

113

1	2		5				8	6
				8	1	7		
5				3	6			1
	3	5	6			9		
				7	5			3
9			8		3			
6	5				4			8
	8		1	6			7	
7		2	3			4		9

				1				4
			8	2			7	3
		5	4		7	6		
6			7	4	3	9	8	
		8	1			7	4	
	7				9	3		
3					1			9
			6		4	2		
	2							6

2	4	7		6			9	1
	3	6	9			7		8
5			1					6
			2				8	
3	2	8				6	5	
7	5		3					
4			6	9			2	
		2			8			3
		5				1		9

	3		1				2	
9	1	8			5	7		4
	4							3
1	7							
	9	6			3		5	1
		5		1	6	9		
	6				7		9	5
	2					3	4	6
		3			2			

DAY

115

	2		5		7	9	8	
			3			1	6	5
8		6			4			
	4							
5				8			2	
	9		6		5		1	8
			7	3			5	
					6			1
			1	5	2			3

6						5	8	
	2				1		4	9
				8			3	
		3	7					8
9	6	7	2			1		3
	8		6				7	
	3						1	4
8	9		1	3				
4			8	5	2			

1	8		3	2	5	6		
		6			9	1		
					4		3	
7							2	
	4		5	9	6		1	
	1							5
	2		4				5	
		8	6			2		
		3	9	5	2		8	1

				2		8		
	2			5	4			
8		9	1		7		4	
		2			1	5	6	4
1	4	8	7			3		
	9		2		8	4		6
			6	4			9	
		6		7				

DAY

116

DAY

117

2					6		8	
6		7	8			2		
	1						4	
8		6	9	2		1		
			6		7			
		1		5	8	9		6
	3						6	
		9			2	3		4
	8		1					2

	2	5				9	6	4
7				6	5	1	2	8
9	6	1			8	7	5	3
1					2			5
			5	1	3			
		9	8			4		
	1		3			5	4	7
			7	5				6
	4					3	9	

			5	4	7		9	2
7			9	1				4
						7		
					5		4	6
	6	4		2		5	3	
5	3		4					
		3						
6				9	3			5
8	9		2	7	4			

	3					1		
8					6	2	3	
1					9	8		
2			5	1	3			
	8	7		2		3	9	1
			9	8	7			2
		9	4					8
	1	3	8					7
		8					1	

DAY

119

	2		6	5	8		1	
	6	9				4	7	
	3			9			6	
	9		2		6		5	
2			8	3	9			1
	4		5		1		9	
				6				
		2				5		
			3	8	5			

			8		2	9		5
	6					8	7	
	3		7					
8	1		5	9	7	3		
		9	1	2	3		6	8
					9		5	
	9	7					8	
4		2	3		8			

	3			6		1	2	
			2		4		8	
6						9		3
		9	3		7			
	8		9		1		5	
			6		2	3		
4		5						1
	9		1		5			
	2	7		9			3	

2		5	9		3		4	
3			1		5			
	4					3		
9		7	6				3	
			5		7			
	5				4	6		9
		2					6	
			3		6			4
	6		4		9	7		8

ANSWER KEY

DAY 1 TOP

5	9	1	4	6	3	7	8	2
3	4	2	7	8	9	6	1	5
7	6	8	1	5	2	4	3	9
9	5	3	8	4	6	1	2	7
2	1	6	3	9	7	5	4	8
8	7	4	5	2	1	3	9	6
6	3	9	2	7	4	8	5	1
1	8	7	9	3	5	2	6	4
4	2	5	6	1	8	9	7	3

DAY 1 BOTTOM

1	6	9	4	5	7	3	2	8
8	2	4	6	1	3	9	5	7
3	7	5	8	2	9	6	4	1
7	9	6	5	8	4	2	1	3
5	8	2	3	6	1	4	7	9
4	3	1	9	7	2	8	6	5
2	1	8	7	9	6	5	3	4
6	5	3	1	4	8	7	9	2
9	4	7	2	3	5	1	8	6

DAY 2 TOP

7	8	6	5	1	9	3	2	4
5	2	1	3	4	6	7	8	9
4	9	3	8	2	7	1	5	6
9	4	7	1	8	2	6	3	5
1	5	8	4	6	3	9	7	2
6	3	2	7	9	5	8	4	1
2	7	9	6	5	8	4	1	3
3	1	5	9	7	4	2	6	8
8	6	4	2	3	1	5	9	7

DAY 2 BOTTOM

6	7	9	3	8	5	4	1	2
4	2	5	1	9	6	7	8	3
8	3	1	4	2	7	5	9	6
7	1	2	5	6	8	3	4	9
5	6	4	9	3	1	8	2	7
3	9	8	7	4	2	1	6	5
9	4	7	2	1	3	6	5	8
2	8	3	6	5	4	9	7	1
1	5	6	8	7	9	2	3	4

DAY 3 TOP

9	4	3	7	1	2	5	6	8
1	7	5	9	8	6	4	3	2
2	8	6	5	3	4	9	7	1
8	6	1	3	4	5	2	9	7
3	5	2	8	7	9	6	1	4
4	9	7	2	6	1	8	5	3
6	3	8	4	9	7	1	2	5
5	1	4	6	2	3	7	8	9
7	2	9	1	5	8	3	4	6

DAY 3 BOTTOM

7	3	2	1	4	5	9	8	6
4	8	5	9	3	6	2	1	7
9	6	1	8	2	7	5	4	3
2	7	6	5	1	4	8	3	9
8	4	9	3	6	2	1	7	5
1	5	3	7	8	9	6	2	4
3	9	8	6	7	1	4	5	2
6	1	4	2	5	3	7	9	8
5	2	7	4	9	8	3	6	1

DAY 4 TOP

2	3	9	1	7	5	6	4	8
4	1	5	6	3	8	7	2	9
7	8	6	4	9	2	3	5	1
8	4	3	7	5	9	1	6	2
5	2	1	3	8	6	4	9	7
6	9	7	2	4	1	5	8	3
9	7	4	5	2	3	8	1	6
3	6	2	8	1	4	9	7	5
1	5	8	9	6	7	2	3	4

DAY 4 BOTTOM

9	8	2	3	1	5	4	7	6
5	3	6	2	7	4	8	9	1
1	7	4	9	6	8	2	5	3
4	9	8	1	5	3	7	6	2
2	1	5	6	8	7	3	4	9
7	6	3	4	2	9	5	1	8
6	2	7	8	4	1	9	3	5
8	5	9	7	3	6	1	2	4
3	4	1	5	9	2	6	8	7

DAY 5 TOP

2	4	5	7	8	9	6	1	3
1	7	8	6	3	5	4	2	9
9	3	6	2	1	4	8	7	5
6	5	1	4	9	7	2	3	8
7	8	2	3	5	6	9	4	1
3	9	4	8	2	1	5	6	7
5	2	3	1	6	8	7	9	4
8	6	7	9	4	3	1	5	2
4	1	9	5	7	2	3	8	6

DAY 5 BOTTOM

6	8	7	2	5	1	4	3	9
2	3	1	7	9	4	5	6	8
4	9	5	3	6	8	1	7	2
3	5	8	4	7	2	6	9	1
7	2	6	5	1	9	8	4	3
9	1	4	8	3	6	2	5	7
8	4	9	6	2	7	3	1	5
1	6	3	9	8	5	7	2	4
5	7	2	1	4	3	9	8	6

DAY 6 TOP

2	6	1	9	8	3	7	4	5
9	4	7	5	6	2	1	8	3
8	3	5	4	7	1	6	2	9
4	5	3	1	2	7	8	9	6
7	2	6	8	3	9	4	5	1
1	9	8	6	4	5	3	7	2
3	7	9	2	1	8	5	6	4
5	1	4	7	9	6	2	3	8
6	8	2	3	5	4	9	1	7

DAY 6 BOTTOM

2	4	9	1	7	6	8	5	3
5	6	8	2	4	3	9	1	7
7	1	3	8	9	5	2	4	6
3	5	7	4	8	9	1	6	2
6	9	1	5	3	2	7	8	4
4	8	2	7	6	1	5	3	9
9	3	5	6	2	8	4	7	1
8	7	6	9	1	4	3	2	5
1	2	4	3	5	7	6	9	8

DAY 7 TOP

9	3	1	5	8	7	2	4	6
6	2	5	9	4	3	8	1	7
8	4	7	2	1	6	3	9	5
2	7	6	4	3	9	1	5	8
4	5	3	1	6	8	7	2	9
1	8	9	7	5	2	4	6	3
3	6	2	8	9	1	5	7	4
7	9	4	3	2	5	6	8	1
5	1	8	6	7	4	9	3	2

DAY 7 BOTTOM

5	4	2	1	8	7	9	3	6
6	7	8	9	3	5	2	4	1
3	9	1	6	4	2	8	5	7
4	2	3	8	7	1	6	9	5
7	6	9	5	2	3	1	8	4
8	1	5	4	9	6	3	7	2
1	3	7	2	5	8	4	6	9
2	8	4	7	6	9	5	1	3
9	5	6	3	1	4	7	2	8

DAY 8 TOP

4	9	1	6	7	3	8	5	2
6	8	2	9	4	5	1	7	3
5	7	3	1	2	8	6	4	9
7	3	4	2	8	1	5	9	6
8	6	9	5	3	4	7	2	1
2	1	5	7	9	6	3	8	4
3	2	7	8	6	9	4	1	5
9	5	6	4	1	7	2	3	8
1	4	8	3	5	2	9	6	7

DAY 8 BOTTOM

6	4	5	1	8	9	2	7	3
1	7	8	4	2	3	6	9	5
3	2	9	5	6	7	4	1	8
5	6	2	9	1	8	3	4	7
7	1	3	2	5	4	8	6	9
9	8	4	7	3	6	1	5	2
4	3	7	6	9	2	5	8	1
8	5	6	3	7	1	9	2	4
2	9	1	8	4	5	7	3	6

DAY 9 TOP

5	7	1	4	8	6	9	2	3
4	2	8	9	5	3	6	1	7
6	3	9	7	2	1	4	5	8
1	8	5	3	6	4	7	9	2
9	4	3	1	7	2	8	6	5
7	6	2	8	9	5	1	3	4
8	5	6	2	4	9	3	7	1
3	9	7	5	1	8	2	4	6
2	1	4	6	3	7	5	8	9

DAY 9 BOTTOM

4	9	7	1	3	5	8	6	2
3	1	6	8	2	4	9	7	5
8	2	5	7	6	9	3	1	4
7	3	1	2	4	8	5	9	6
9	5	2	3	7	6	4	8	1
6	4	8	9	5	1	7	2	3
2	7	4	6	8	3	1	5	9
5	6	9	4	1	7	2	3	8
1	8	3	5	9	2	6	4	7

DAY **10** TOP

8	9	2	3	4	5	7	1	6
3	1	5	7	2	6	8	9	4
6	4	7	9	8	1	5	3	2
2	7	1	8	3	4	9	6	5
5	3	9	6	7	2	4	8	1
4	6	8	5	1	9	2	7	3
7	5	3	4	6	8	1	2	9
1	8	4	2	9	3	6	5	7
9	2	6	1	5	7	3	4	8

DAY **10** BOTTOM

3	6	9	8	7	4	2	1	5
1	2	7	6	3	5	4	8	9
4	5	8	1	2	9	6	7	3
2	8	1	3	5	6	9	4	7
5	7	6	4	9	1	8	3	2
9	4	3	2	8	7	5	6	1
7	9	4	5	1	8	3	2	6
8	3	5	7	6	2	1	9	4
6	1	2	9	4	3	7	5	8

DAY **11** TOP

8	5	2	6	3	9	1	7	4
3	4	1	8	7	2	9	6	5
9	7	6	5	1	4	8	3	2
5	9	3	7	2	1	4	8	6
1	6	7	4	5	8	3	2	9
4	2	8	3	9	6	7	5	1
7	1	5	2	4	3	6	9	8
6	3	4	9	8	5	2	1	7
2	8	9	1	6	7	5	4	3

DAY **11** BOTTOM

9	3	1	2	8	7	4	5	6
4	6	2	9	3	5	8	1	7
7	5	8	6	4	1	2	9	3
1	7	9	8	6	2	3	4	5
6	8	4	5	9	3	7	2	1
5	2	3	1	7	4	9	6	8
8	9	5	3	2	6	1	7	4
3	4	6	7	1	9	5	8	2
2	1	7	4	5	8	6	3	9

DAY **12** TOP

1	7	4	5	6	8	2	9	3
5	3	6	7	9	2	4	1	8
2	9	8	4	1	3	6	7	5
3	5	2	9	4	7	8	6	1
4	1	9	8	5	6	3	2	7
8	6	7	3	2	1	9	5	4
9	8	5	6	7	4	1	3	2
7	2	3	1	8	9	5	4	6
6	4	1	2	3	5	7	8	9

DAY **12** BOTTOM

7	9	4	8	5	6	1	2	3
5	3	8	1	2	4	6	7	9
1	6	2	3	9	7	4	8	5
6	1	7	9	8	3	5	4	2
8	5	9	4	1	2	3	6	7
4	2	3	6	7	5	8	9	1
2	8	5	7	6	1	9	3	4
3	7	6	5	4	9	2	1	8
9	4	1	2	3	8	7	5	6

DAY 13 TOP

2	3	8	9	5	1	6	4	7
9	1	4	3	7	6	5	2	8
6	5	7	4	2	8	1	3	9
4	8	5	6	3	9	2	7	1
3	2	9	5	1	7	8	6	4
7	6	1	8	4	2	9	5	3
8	7	6	2	9	3	4	1	5
5	9	3	1	6	4	7	8	2
1	4	2	7	8	5	3	9	6

DAY 13 BOTTOM

3	1	6	5	9	2	8	7	4
2	4	7	3	1	8	5	6	9
9	5	8	6	7	4	2	1	3
4	8	5	7	3	6	1	9	2
7	3	1	2	5	9	6	4	8
6	2	9	8	4	1	3	5	7
8	6	4	1	2	7	9	3	5
5	7	2	9	6	3	4	8	1
1	9	3	4	8	5	7	2	6

DAY 14 TOP

2	6	8	7	5	4	3	9	1
3	5	9	6	1	2	4	8	7
7	4	1	8	3	9	6	2	5
6	9	2	5	4	7	1	3	8
1	3	7	9	8	6	5	4	2
5	8	4	1	2	3	9	7	6
9	7	5	4	6	8	2	1	3
4	2	6	3	7	1	8	5	9
8	1	3	2	9	5	7	6	4

DAY 14 BOTTOM

8	6	1	9	7	4	5	2	3
7	3	9	6	2	5	8	1	4
4	2	5	8	1	3	7	6	9
5	9	3	2	8	6	4	7	1
2	1	8	4	9	7	3	5	6
6	7	4	3	5	1	2	9	8
3	5	6	1	4	2	9	8	7
9	4	2	7	6	8	1	3	5
1	8	7	5	3	9	6	4	2

DAY 15 TOP

9	4	2	6	3	5	8	7	1
1	3	7	4	8	9	6	2	5
6	5	8	2	1	7	3	9	4
8	1	6	7	2	4	9	5	3
3	2	5	8	9	1	4	6	7
4	7	9	3	5	6	2	1	8
5	9	4	1	6	8	7	3	2
7	6	3	5	4	2	1	8	9
2	8	1	9	7	3	5	4	6

DAY 15 BOTTOM

6	8	7	2	1	4	3	9	5
3	2	5	6	9	7	4	8	1
1	4	9	5	8	3	2	6	7
8	6	1	4	2	5	7	3	9
5	3	4	9	7	6	1	2	8
9	7	2	8	3	1	6	5	4
7	5	3	1	6	8	9	4	2
4	9	6	7	5	2	8	1	3
2	1	8	3	4	9	5	7	6

DAY **16** TOP

5	6	2	7	3	9	1	4	8
7	4	8	6	1	5	9	2	3
1	9	3	2	4	8	5	7	6
8	2	1	9	7	6	4	3	5
4	7	6	1	5	3	8	9	2
9	3	5	8	2	4	6	1	7
3	8	9	4	6	7	2	5	1
6	1	7	5	9	2	3	8	4
2	5	4	3	8	1	7	6	9

DAY **16** BOTTOM

7	1	5	6	3	4	9	8	2
6	9	3	1	8	2	4	7	5
2	8	4	5	7	9	6	1	3
9	6	2	8	5	1	7	3	4
4	3	1	9	2	7	5	6	8
5	7	8	4	6	3	1	2	9
1	2	9	7	4	8	3	5	6
8	4	6	3	1	5	2	9	7
3	5	7	2	9	6	8	4	1

DAY **17** TOP

7	1	4	6	5	9	8	3	2
2	5	9	3	4	8	6	7	1
3	8	6	1	2	7	9	5	4
6	2	7	4	8	5	1	9	3
5	4	1	9	6	3	7	2	8
9	3	8	2	7	1	5	4	6
4	6	5	8	9	2	3	1	7
1	9	2	7	3	6	4	8	5
8	7	3	5	1	4	2	6	9

DAY **17** BOTTOM

7	2	3	5	1	4	6	9	8
9	8	5	6	3	7	1	4	2
1	6	4	9	2	8	5	3	7
3	7	8	4	5	1	2	6	9
6	1	9	2	8	3	7	5	4
5	4	2	7	6	9	3	8	1
4	3	7	1	9	6	8	2	5
8	5	1	3	4	2	9	7	6
2	9	6	8	7	5	4	1	3

DAY **18** TOP

9	4	6	1	3	2	5	7	8
8	7	2	9	5	4	6	1	3
5	3	1	7	6	8	4	2	9
2	5	9	3	7	6	8	4	1
1	8	7	4	9	5	2	3	6
3	6	4	8	2	1	7	9	5
7	9	8	6	4	3	1	5	2
6	2	3	5	1	7	9	8	4
4	1	5	2	8	9	3	6	7

DAY **18** BOTTOM

3	4	2	5	1	9	6	8	7
8	9	5	6	7	2	1	4	3
7	6	1	3	4	8	5	2	9
9	2	4	7	5	1	3	6	8
1	8	3	9	2	6	4	7	5
5	7	6	4	8	3	9	1	2
6	3	7	2	9	4	8	5	1
2	1	9	8	6	5	7	3	4
4	5	8	1	3	7	2	9	6

DAY **19** TOP

8	7	5	2	6	3	9	4	1
9	4	3	1	8	7	2	6	5
1	2	6	5	4	9	8	3	7
3	1	8	4	2	5	6	7	9
2	5	7	3	9	6	1	8	4
6	9	4	8	7	1	5	2	3
5	3	2	6	1	4	7	9	8
4	8	9	7	5	2	3	1	6
7	6	1	9	3	8	4	5	2

DAY **19** BOTTOM

6	5	2	3	4	8	9	7	1
7	4	1	9	6	5	8	2	3
8	9	3	7	2	1	5	4	6
2	1	9	4	8	6	7	3	5
4	7	6	5	9	3	1	8	2
3	8	5	2	1	7	6	9	4
1	3	8	6	7	4	2	5	9
5	2	7	1	3	9	4	6	8
9	6	4	8	5	2	3	1	7

DAY **20** TOP

3	4	9	2	7	6	5	8	1
7	1	2	4	8	5	3	9	6
6	8	5	3	9	1	7	4	2
4	7	6	1	5	9	8	2	3
8	9	3	6	2	7	1	5	4
5	2	1	8	3	4	6	7	9
2	6	8	7	4	3	9	1	5
1	5	4	9	6	8	2	3	7
9	3	7	5	1	2	4	6	8

DAY **20** BOTTOM

7	3	2	9	8	1	4	5	6
5	8	6	3	7	4	9	2	1
9	4	1	6	2	5	3	8	7
8	5	7	1	4	3	6	9	2
2	6	4	7	9	8	5	1	3
1	9	3	5	6	2	7	4	8
3	2	5	4	1	6	8	7	9
6	1	9	8	5	7	2	3	4
4	7	8	2	3	9	1	6	5

DAY **21** TOP

5	1	7	2	4	8	6	3	9
3	8	9	6	1	7	4	2	5
2	6	4	9	3	5	1	7	8
7	9	8	1	2	6	3	5	4
1	4	2	5	9	3	7	8	6
6	3	5	8	7	4	9	1	2
8	7	6	4	5	1	2	9	3
9	5	1	3	6	2	8	4	7
4	2	3	7	8	9	5	6	1

DAY **21** BOTTOM

1	5	7	4	2	3	6	8	9
8	4	9	1	7	6	5	2	3
3	2	6	9	8	5	1	4	7
9	8	2	6	1	7	4	3	5
5	1	3	2	9	4	8	7	6
6	7	4	5	3	8	9	1	2
2	6	1	7	4	9	3	5	8
4	3	5	8	6	2	7	9	1
7	9	8	3	5	1	2	6	4

DAY **22** TOP

6	5	3	4	2	1	7	8	9
2	9	4	8	3	7	5	1	6
8	7	1	5	9	6	2	4	3
3	2	6	7	1	4	9	5	8
7	4	8	6	5	9	3	2	1
9	1	5	2	8	3	4	6	7
1	3	2	9	4	8	6	7	5
5	8	7	3	6	2	1	9	4
4	6	9	1	7	5	8	3	2

DAY **22** BOTTOM

1	7	4	9	3	5	8	6	2
5	9	8	4	6	2	3	1	7
3	6	2	1	8	7	9	5	4
4	3	5	8	7	6	1	2	9
6	8	9	2	4	1	7	3	5
2	1	7	3	5	9	4	8	6
9	2	3	5	1	4	6	7	8
7	5	1	6	9	8	2	4	3
8	4	6	7	2	3	5	9	1

DAY **23** TOP

2	8	6	1	5	3	4	9	7
3	9	4	7	2	6	8	1	5
7	1	5	4	9	8	3	6	2
9	7	8	3	6	5	1	2	4
5	2	1	8	4	9	7	3	6
4	6	3	2	1	7	5	8	9
6	3	7	5	8	2	9	4	1
1	5	9	6	3	4	2	7	8
8	4	2	9	7	1	6	5	3

DAY **23** BOTTOM

1	7	2	3	5	9	8	6	4
6	5	4	8	2	7	9	3	1
3	8	9	4	1	6	7	2	5
7	9	3	1	6	4	2	5	8
2	4	5	7	8	3	6	1	9
8	6	1	5	9	2	4	7	3
5	2	8	9	7	1	3	4	6
9	3	6	2	4	5	1	8	7
4	1	7	6	3	8	5	9	2

DAY **24** TOP

4	6	1	5	8	7	9	3	2
3	7	9	2	6	4	5	1	8
8	2	5	3	9	1	4	7	6
7	3	4	6	5	2	1	8	9
5	8	6	1	3	9	7	2	4
1	9	2	7	4	8	6	5	3
6	1	3	9	2	5	8	4	7
2	4	7	8	1	6	3	9	5
9	5	8	4	7	3	2	6	1

DAY **24** BOTTOM

1	7	9	5	3	6	4	8	2
4	5	2	9	8	7	3	6	1
3	8	6	1	2	4	9	7	5
9	2	5	8	7	3	1	4	6
7	6	3	4	1	2	5	9	8
8	4	1	6	5	9	7	2	3
2	1	7	3	4	8	6	5	9
5	9	8	7	6	1	2	3	4
6	3	4	2	9	5	8	1	7

DAY 25 TOP

9	1	5	8	4	3	7	6	2
2	4	7	1	5	6	9	8	3
3	6	8	2	9	7	4	1	5
6	2	9	5	8	4	3	7	1
1	8	3	7	2	9	6	5	4
5	7	4	3	6	1	8	2	9
7	9	1	6	3	2	5	4	8
4	5	6	9	1	8	2	3	7
8	3	2	4	7	5	1	9	6

DAY 25 BOTTOM

4	6	3	2	7	9	1	8	5
1	5	7	8	6	3	2	9	4
2	9	8	5	4	1	6	3	7
6	7	5	3	8	2	4	1	9
3	2	9	4	1	5	7	6	8
8	4	1	7	9	6	3	5	2
7	8	6	9	3	4	5	2	1
9	1	2	6	5	7	8	4	3
5	3	4	1	2	8	9	7	6

DAY 26 TOP

3	9	6	7	5	2	8	1	4
7	5	8	1	4	3	6	2	9
4	1	2	8	6	9	5	3	7
8	3	4	2	9	5	1	7	6
9	2	1	6	7	8	4	5	3
6	7	5	3	1	4	9	8	2
5	4	7	9	2	1	3	6	8
1	6	3	4	8	7	2	9	5
2	8	9	5	3	6	7	4	1

DAY 26 BOTTOM

2	5	4	9	8	1	6	7	3
9	7	3	4	6	2	1	8	5
6	8	1	5	3	7	2	4	9
1	2	8	7	5	3	4	9	6
7	6	5	1	9	4	3	2	8
3	4	9	6	2	8	7	5	1
4	9	6	2	1	5	8	3	7
5	3	2	8	7	6	9	1	4
8	1	7	3	4	9	5	6	2

DAY 27 TOP

9	3	8	5	6	7	1	4	2
7	6	4	2	1	8	9	5	3
2	5	1	4	9	3	8	7	6
4	2	7	8	3	1	6	9	5
5	8	6	9	2	4	7	3	1
3	1	9	6	7	5	2	8	4
1	7	5	3	8	6	4	2	9
6	9	3	7	4	2	5	1	8
8	4	2	1	5	9	3	6	7

DAY 27 BOTTOM

3	8	7	1	2	5	6	9	4
1	4	2	6	9	3	5	7	8
9	5	6	4	8	7	3	1	2
7	6	5	8	1	4	2	3	9
4	9	3	5	7	2	1	8	6
8	2	1	3	6	9	4	5	7
6	3	8	9	4	1	7	2	5
5	7	4	2	3	8	9	6	1
2	1	9	7	5	6	8	4	3

DAY **28** TOP

8	2	1	6	9	5	3	7	4
3	7	9	4	8	1	2	6	5
5	6	4	3	2	7	9	8	1
2	4	7	9	5	8	6	1	3
9	8	3	1	4	6	5	2	7
6	1	5	2	7	3	4	9	8
7	5	6	8	3	9	1	4	2
4	9	8	5	1	2	7	3	6
1	3	2	7	6	4	8	5	9

DAY **28** BOTTOM

1	4	6	3	7	5	9	8	2
5	3	2	8	4	9	1	7	6
8	9	7	2	1	6	3	4	5
3	6	5	4	9	8	7	2	1
4	2	9	7	5	1	6	3	8
7	1	8	6	2	3	4	5	9
9	5	3	1	8	7	2	6	4
2	7	1	5	6	4	8	9	3
6	8	4	9	3	2	5	1	7

DAY **29** TOP

8	3	6	4	2	5	7	9	1
4	1	9	7	8	3	2	6	5
5	2	7	1	6	9	4	3	8
9	8	3	5	4	2	6	1	7
1	7	2	6	9	8	3	5	4
6	4	5	3	7	1	8	2	9
2	9	4	8	5	6	1	7	3
3	5	8	2	1	7	9	4	6
7	6	1	9	3	4	5	8	2

DAY **29** BOTTOM

2	5	6	8	4	7	3	9	1
8	1	9	5	2	3	4	7	6
3	4	7	6	9	1	2	5	8
1	8	5	3	7	2	9	6	4
9	6	3	4	5	8	1	2	7
4	7	2	9	1	6	5	8	3
7	9	4	1	8	5	6	3	2
6	2	1	7	3	9	8	4	5
5	3	8	2	6	4	7	1	9

DAY **30** TOP

9	1	5	3	4	6	7	8	2
4	3	2	5	8	7	9	1	6
6	7	8	9	1	2	4	5	3
7	2	3	8	9	5	6	4	1
5	6	4	2	7	1	8	3	9
8	9	1	6	3	4	2	7	5
3	8	6	4	5	9	1	2	7
2	5	7	1	6	8	3	9	4
1	4	9	7	2	3	5	6	8

DAY **30** BOTTOM

6	7	8	3	2	5	9	1	4
9	5	2	1	4	8	7	6	3
1	4	3	7	6	9	8	5	2
5	8	7	2	1	3	4	9	6
3	9	6	8	5	4	1	2	7
2	1	4	9	7	6	3	8	5
8	3	5	4	9	2	6	7	1
7	2	9	6	3	1	5	4	8
4	6	1	5	8	7	2	3	9

DAY 31 TOP

6	5	7	8	9	4	3	1	2
9	3	2	1	5	6	7	4	8
1	4	8	3	7	2	9	6	5
3	9	5	6	1	8	4	2	7
2	7	1	9	4	5	8	3	6
4	8	6	7	2	3	1	5	9
5	1	9	4	6	7	2	8	3
7	6	3	2	8	1	5	9	4
8	2	4	5	3	9	6	7	1

DAY 31 BOTTOM

4	8	6	5	1	7	9	3	2
3	7	1	6	9	2	5	4	8
5	9	2	8	3	4	6	7	1
9	5	8	3	4	1	7	2	6
7	1	4	2	6	9	3	8	5
2	6	3	7	5	8	1	9	4
1	2	7	9	8	5	4	6	3
6	4	9	1	2	3	8	5	7
8	3	5	4	7	6	2	1	9

DAY 32 TOP

7	3	9	6	2	1	8	4	5
2	8	1	9	4	5	3	7	6
5	6	4	7	3	8	2	1	9
6	5	7	8	1	3	4	9	2
1	2	3	4	5	9	7	6	8
9	4	8	2	7	6	5	3	1
3	9	5	1	8	4	6	2	7
4	1	2	5	6	7	9	8	3
8	7	6	3	9	2	1	5	4

DAY 32 BOTTOM

4	5	3	2	9	8	7	6	1
1	8	9	6	5	7	2	4	3
6	2	7	3	1	4	8	9	5
2	4	1	5	7	9	3	8	6
8	7	6	1	4	3	9	5	2
9	3	5	8	2	6	4	1	7
3	1	8	4	6	2	5	7	9
5	9	2	7	8	1	6	3	4
7	6	4	9	3	5	1	2	8

DAY 33 TOP

5	2	4	1	8	3	7	6	9
6	9	3	7	5	2	4	1	8
7	1	8	4	9	6	2	5	3
1	4	2	3	6	8	9	7	5
8	6	9	5	2	7	3	4	1
3	7	5	9	1	4	6	8	2
4	3	1	8	7	9	5	2	6
2	8	7	6	3	5	1	9	4
9	5	6	2	4	1	8	3	7

DAY 33 BOTTOM

4	1	8	6	7	9	5	2	3
3	7	5	8	2	1	6	9	4
2	6	9	3	5	4	1	7	8
6	4	7	5	8	2	3	1	9
8	3	1	9	6	7	2	4	5
9	5	2	1	4	3	7	8	6
1	8	6	2	9	5	4	3	7
5	2	4	7	3	8	9	6	1
7	9	3	4	1	6	8	5	2

DAY **34** TOP

8	6	3	9	7	4	5	2	1
1	4	2	5	6	8	7	9	3
7	5	9	1	2	3	6	8	4
9	7	6	3	1	2	8	4	5
4	1	8	6	9	5	2	3	7
3	2	5	4	8	7	1	6	9
6	3	1	2	5	9	4	7	8
5	8	4	7	3	6	9	1	2
2	9	7	8	4	1	3	5	6

DAY **34** BOTTOM

5	2	9	4	6	7	8	3	1
6	3	4	1	8	5	7	2	9
1	7	8	2	3	9	5	4	6
2	4	6	8	5	1	3	9	7
3	1	5	9	7	4	6	8	2
9	8	7	6	2	3	4	1	5
8	5	2	3	1	6	9	7	4
7	9	3	5	4	2	1	6	8
4	6	1	7	9	8	2	5	3

DAY **35** TOP

5	7	8	9	2	3	6	1	4
6	3	1	7	4	5	8	2	9
9	4	2	1	6	8	3	5	7
3	6	5	2	7	4	9	8	1
2	9	7	6	8	1	4	3	5
8	1	4	5	3	9	2	7	6
7	8	6	4	1	2	5	9	3
4	5	3	8	9	7	1	6	2
1	2	9	3	5	6	7	4	8

DAY **35** BOTTOM

8	1	3	4	2	5	9	7	6
6	4	5	9	7	8	3	1	2
9	2	7	3	6	1	8	4	5
3	9	1	5	4	2	7	6	8
5	7	6	1	8	9	4	2	3
4	8	2	7	3	6	5	9	1
7	6	8	2	9	3	1	5	4
2	5	9	8	1	4	6	3	7
1	3	4	6	5	7	2	8	9

DAY **36** TOP

6	8	3	4	5	1	9	2	7
1	9	7	6	2	3	5	4	8
2	5	4	9	7	8	3	6	1
9	2	8	3	1	4	6	7	5
7	3	1	5	6	2	8	9	4
4	6	5	7	8	9	1	3	2
3	1	2	8	4	6	7	5	9
5	4	9	1	3	7	2	8	6
8	7	6	2	9	5	4	1	3

DAY **36** BOTTOM

4	7	8	3	1	6	9	5	2
5	3	9	2	7	8	6	4	1
2	6	1	9	4	5	7	8	3
1	9	2	4	5	7	3	6	8
8	5	7	6	3	1	2	9	4
6	4	3	8	2	9	5	1	7
9	2	4	5	8	3	1	7	6
3	1	5	7	6	4	8	2	9
7	8	6	1	9	2	4	3	5

DAY **37** TOP

8	4	7	3	2	9	6	5	1
5	1	3	8	4	6	7	2	9
9	6	2	1	5	7	8	3	4
6	3	4	7	9	5	1	8	2
7	9	8	2	1	3	5	4	6
1	2	5	6	8	4	3	9	7
3	7	9	5	6	2	4	1	8
2	8	6	4	3	1	9	7	5
4	5	1	9	7	8	2	6	3

DAY **37** BOTTOM

2	6	8	1	4	7	5	9	3
9	3	1	2	5	8	7	6	4
7	4	5	6	9	3	8	1	2
4	8	7	3	6	9	2	5	1
6	5	3	7	1	2	9	4	8
1	2	9	5	8	4	3	7	6
3	7	4	9	2	1	6	8	5
5	1	2	8	7	6	4	3	9
8	9	6	4	3	5	1	2	7

DAY **38** TOP

9	7	8	5	2	4	1	6	3
5	6	2	3	1	8	4	7	9
1	4	3	7	6	9	8	5	2
4	1	9	2	7	6	5	3	8
8	3	6	9	4	5	2	1	7
2	5	7	1	8	3	6	9	4
6	9	4	8	5	7	3	2	1
7	8	1	6	3	2	9	4	5
3	2	5	4	9	1	7	8	6

DAY **38** BOTTOM

2	9	5	3	4	1	6	8	7
6	3	4	8	7	9	2	1	5
1	7	8	6	2	5	4	9	3
4	2	3	9	8	7	5	6	1
8	5	1	2	3	6	7	4	9
7	6	9	1	5	4	8	3	2
5	1	6	4	9	2	3	7	8
9	8	2	7	6	3	1	5	4
3	4	7	5	1	8	9	2	6

DAY **39** TOP

2	6	3	1	5	8	4	9	7
5	4	8	7	3	9	2	1	6
9	7	1	2	4	6	3	8	5
3	5	2	8	1	7	9	6	4
7	8	9	4	6	2	1	5	3
4	1	6	3	9	5	7	2	8
8	2	4	6	7	1	5	3	9
6	9	7	5	2	3	8	4	1
1	3	5	9	8	4	6	7	2

DAY **39** BOTTOM

2	1	8	3	4	9	5	6	7
7	5	4	1	6	8	3	9	2
6	9	3	7	5	2	8	4	1
9	6	5	8	2	7	4	1	3
3	4	2	5	1	6	7	8	9
1	8	7	9	3	4	6	2	5
5	2	6	4	7	1	9	3	8
8	3	1	6	9	5	2	7	4
4	7	9	2	8	3	1	5	6

DAY **40** TOP

9	7	5	6	3	4	2	1	8
3	2	8	1	5	9	6	4	7
1	6	4	8	7	2	5	9	3
8	5	3	9	1	6	7	2	4
4	9	6	7	2	8	3	5	1
2	1	7	3	4	5	9	8	6
7	8	9	2	6	1	4	3	5
5	3	1	4	9	7	8	6	2
6	4	2	5	8	3	1	7	9

DAY **40** BOTTOM

3	4	6	9	1	8	2	7	5
1	5	9	2	4	7	3	6	8
7	2	8	3	6	5	4	9	1
5	1	7	6	2	4	9	8	3
6	3	2	1	8	9	7	5	4
8	9	4	7	5	3	1	2	6
2	8	1	4	9	6	5	3	7
4	7	5	8	3	2	6	1	9
9	6	3	5	7	1	8	4	2

DAY **41** TOP

7	8	5	9	4	6	3	2	1
2	4	3	8	1	5	6	7	9
1	6	9	7	3	2	8	5	4
9	3	1	2	7	8	4	6	5
6	7	4	1	5	3	2	9	8
5	2	8	4	6	9	1	3	7
8	5	6	3	9	1	7	4	2
3	1	7	5	2	4	9	8	6
4	9	2	6	8	7	5	1	3

DAY **41** BOTTOM

3	6	1	5	4	9	2	7	8
5	2	8	6	1	7	4	3	9
4	9	7	8	2	3	5	6	1
6	3	9	4	8	5	1	2	7
1	7	5	3	9	2	6	8	4
2	8	4	7	6	1	9	5	3
8	1	2	9	3	6	7	4	5
7	4	6	1	5	8	3	9	2
9	5	3	2	7	4	8	1	6

DAY **42** TOP

6	5	4	8	9	1	7	3	2
7	9	3	6	2	5	8	4	1
1	8	2	3	7	4	5	6	9
9	4	8	7	6	2	3	1	5
2	7	1	5	4	3	9	8	6
5	3	6	1	8	9	2	7	4
8	6	5	2	1	7	4	9	3
4	2	7	9	3	6	1	5	8
3	1	9	4	5	8	6	2	7

DAY **42** BOTTOM

5	6	9	4	7	3	2	1	8
7	3	1	8	2	5	4	6	9
4	2	8	6	9	1	7	5	3
9	1	4	5	3	2	6	8	7
2	8	7	1	6	9	3	4	5
6	5	3	7	8	4	9	2	1
3	4	5	2	1	7	8	9	6
1	7	6	9	4	8	5	3	2
8	9	2	3	5	6	1	7	4

DAY 43 TOP

5	9	6	3	7	1	2	4	8
3	7	2	4	5	8	9	1	6
1	4	8	6	2	9	5	3	7
4	8	3	9	6	2	1	7	5
6	2	1	7	4	5	8	9	3
9	5	7	8	1	3	4	6	2
2	3	9	1	8	6	7	5	4
8	6	4	5	9	7	3	2	1
7	1	5	2	3	4	6	8	9

DAY 43 BOTTOM

5	1	9	7	4	3	6	8	2
8	4	6	1	2	5	9	3	7
7	3	2	9	8	6	5	1	4
1	6	5	4	3	7	8	2	9
4	9	3	2	5	8	7	6	1
2	8	7	6	9	1	3	4	5
3	5	1	8	7	2	4	9	6
9	2	8	5	6	4	1	7	3
6	7	4	3	1	9	2	5	8

DAY 44 TOP

8	6	3	1	5	4	7	9	2
5	4	9	7	2	3	8	1	6
1	7	2	9	6	8	5	4	3
3	1	5	6	7	9	2	8	4
2	8	6	5	4	1	9	3	7
7	9	4	3	8	2	1	6	5
6	5	8	4	9	7	3	2	1
4	2	1	8	3	5	6	7	9
9	3	7	2	1	6	4	5	8

DAY 44 BOTTOM

5	6	8	2	9	3	1	4	7
3	2	1	4	7	6	5	9	8
4	9	7	8	5	1	6	2	3
6	5	3	7	2	9	4	8	1
1	8	9	3	6	4	7	5	2
7	4	2	5	1	8	9	3	6
2	1	5	9	8	7	3	6	4
9	3	6	1	4	2	8	7	5
8	7	4	6	3	5	2	1	9

DAY 45 TOP

2	3	9	5	4	8	1	7	6
7	5	8	1	3	6	4	2	9
4	6	1	7	9	2	3	8	5
1	9	4	2	7	5	8	6	3
8	2	5	3	6	1	7	9	4
6	7	3	4	8	9	5	1	2
5	8	6	9	1	4	2	3	7
9	4	7	8	2	3	6	5	1
3	1	2	6	5	7	9	4	8

DAY 45 BOTTOM

8	7	6	1	3	4	2	5	9
2	4	5	8	9	6	3	1	7
9	1	3	5	7	2	8	4	6
1	5	2	6	8	9	7	3	4
4	8	9	3	5	7	1	6	2
3	6	7	2	4	1	9	8	5
6	9	1	4	2	8	5	7	3
7	3	4	9	1	5	6	2	8
5	2	8	7	6	3	4	9	1

DAY 46 TOP

5	9	4	7	6	1	2	3	8
2	1	8	9	4	3	5	7	6
3	6	7	5	2	8	4	1	9
7	8	3	2	9	5	6	4	1
4	2	9	1	7	6	8	5	3
6	5	1	3	8	4	7	9	2
1	7	2	6	5	9	3	8	4
8	3	5	4	1	2	9	6	7
9	4	6	8	3	7	1	2	5

DAY 46 BOTTOM

3	4	2	7	1	6	5	8	9
6	7	8	2	9	5	4	1	3
1	9	5	3	8	4	2	7	6
2	1	6	8	4	9	7	3	5
9	8	7	5	2	3	6	4	1
5	3	4	1	6	7	9	2	8
4	6	3	9	7	8	1	5	2
7	5	1	6	3	2	8	9	4
8	2	9	4	5	1	3	6	7

DAY 47 TOP

5	2	4	6	7	9	3	1	8
3	9	8	5	2	1	6	7	4
7	6	1	8	4	3	2	9	5
9	4	6	3	1	5	8	2	7
1	3	7	2	8	4	9	5	6
2	8	5	7	9	6	1	4	3
4	1	3	9	5	8	7	6	2
6	5	2	1	3	7	4	8	9
8	7	9	4	6	2	5	3	1

DAY 47 BOTTOM

3	4	2	9	6	5	8	7	1
9	7	5	1	3	8	2	4	6
8	6	1	7	2	4	9	5	3
7	8	3	6	5	2	4	1	9
2	1	6	4	7	9	3	8	5
4	5	9	3	8	1	6	2	7
5	9	8	2	1	6	7	3	4
6	2	7	5	4	3	1	9	8
1	3	4	8	9	7	5	6	2

DAY 48 TOP

9	2	5	3	8	7	6	1	4
1	8	4	6	5	2	7	3	9
3	6	7	4	9	1	2	5	8
2	7	3	9	1	4	5	8	6
5	4	8	2	6	3	1	9	7
6	1	9	8	7	5	4	2	3
8	5	6	7	2	9	3	4	1
7	3	1	5	4	8	9	6	2
4	9	2	1	3	6	8	7	5

DAY 48 BOTTOM

3	4	5	6	8	2	7	9	1
6	1	8	5	7	9	2	4	3
7	9	2	1	4	3	8	6	5
1	2	4	7	6	5	3	8	9
9	8	7	4	3	1	6	5	2
5	6	3	2	9	8	4	1	7
8	3	1	9	2	4	5	7	6
4	7	9	3	5	6	1	2	8
2	5	6	8	1	7	9	3	4

DAY 49 TOP

7	2	4	8	3	1	6	5	9
1	3	9	5	4	6	2	7	8
8	5	6	9	7	2	4	1	3
6	7	2	1	8	3	5	9	4
4	8	1	2	9	5	3	6	7
5	9	3	4	6	7	1	8	2
2	1	7	3	5	9	8	4	6
3	6	8	7	1	4	9	2	5
9	4	5	6	2	8	7	3	1

DAY 49 BOTTOM

3	5	1	4	7	2	8	6	9
7	6	2	3	8	9	4	5	1
9	4	8	1	6	5	3	2	7
1	8	4	6	5	7	2	9	3
5	2	7	9	3	8	6	1	4
6	9	3	2	4	1	7	8	5
4	1	6	5	2	3	9	7	8
2	7	9	8	1	4	5	3	6
8	3	5	7	9	6	1	4	2

DAY 50 TOP

7	3	2	1	8	6	4	5	9
1	5	4	3	9	2	8	6	7
6	8	9	4	7	5	2	1	3
2	9	1	7	5	8	6	3	4
5	4	6	9	2	3	7	8	1
3	7	8	6	1	4	9	2	5
9	6	3	2	4	1	5	7	8
4	2	5	8	3	7	1	9	6
8	1	7	5	6	9	3	4	2

DAY 50 BOTTOM

6	4	2	8	3	7	1	5	9
3	5	9	6	1	4	8	7	2
7	8	1	2	5	9	3	6	4
9	6	4	3	7	8	2	1	5
5	1	8	9	2	6	4	3	7
2	3	7	1	4	5	6	9	8
8	9	3	7	6	2	5	4	1
4	7	6	5	8	1	9	2	3
1	2	5	4	9	3	7	8	6

DAY 51 TOP

1	5	7	3	8	9	2	6	4
6	9	4	2	7	1	5	8	3
3	8	2	5	6	4	1	9	7
2	7	9	6	3	5	4	1	8
4	6	8	7	1	2	9	3	5
5	1	3	9	4	8	6	7	2
8	2	1	4	9	7	3	5	6
9	3	5	8	2	6	7	4	1
7	4	6	1	5	3	8	2	9

DAY 51 BOTTOM

9	5	8	1	2	3	6	7	4
6	1	7	9	4	8	3	2	5
2	3	4	7	5	6	1	8	9
5	9	2	3	1	7	8	4	6
7	6	3	4	8	9	2	5	1
4	8	1	5	6	2	9	3	7
1	2	5	6	3	4	7	9	8
8	7	6	2	9	5	4	1	3
3	4	9	8	7	1	5	6	2

DAY 52 TOP

4	6	2	7	1	5	9	3	8
3	5	1	6	9	8	2	4	7
7	9	8	2	4	3	5	1	6
8	2	7	3	5	4	6	9	1
1	4	5	9	8	6	7	2	3
9	3	6	1	2	7	4	8	5
2	7	3	4	6	1	8	5	9
6	8	4	5	3	9	1	7	2
5	1	9	8	7	2	3	6	4

DAY 52 BOTTOM

1	3	2	4	5	8	6	9	7
8	7	5	6	9	2	3	4	1
9	6	4	1	3	7	8	5	2
6	5	9	7	1	3	2	8	4
7	4	3	2	8	9	5	1	6
2	8	1	5	6	4	7	3	9
4	2	8	9	7	5	1	6	3
5	1	7	3	4	6	9	2	8
3	9	6	8	2	1	4	7	5

DAY 53 TOP

6	5	9	3	2	7	1	8	4
1	7	8	5	4	9	2	6	3
2	4	3	1	8	6	9	7	5
5	3	2	8	6	1	7	4	9
9	8	7	2	5	4	6	3	1
4	1	6	9	7	3	5	2	8
7	9	4	6	3	5	8	1	2
8	6	5	4	1	2	3	9	7
3	2	1	7	9	8	4	5	6

DAY 53 BOTTOM

1	7	9	6	8	5	2	3	4
3	5	4	7	9	2	6	1	8
2	8	6	1	3	4	5	9	7
6	4	2	3	7	1	9	8	5
8	1	5	2	4	9	7	6	3
9	3	7	5	6	8	1	4	2
5	6	3	8	1	7	4	2	9
7	9	8	4	2	6	3	5	1
4	2	1	9	5	3	8	7	6

DAY 54 TOP

1	4	7	5	8	2	9	6	3
9	2	3	4	7	6	1	5	8
6	8	5	1	9	3	7	2	4
8	1	2	6	3	7	5	4	9
5	3	6	9	4	1	8	7	2
7	9	4	2	5	8	3	1	6
4	7	8	3	6	5	2	9	1
3	6	1	7	2	9	4	8	5
2	5	9	8	1	4	6	3	7

DAY 54 BOTTOM

9	2	7	4	3	5	1	8	6
6	8	4	9	1	7	5	3	2
5	3	1	2	6	8	7	9	4
3	7	5	8	2	6	9	4	1
2	1	8	5	4	9	6	7	3
4	9	6	1	7	3	8	2	5
1	4	9	7	5	2	3	6	8
8	5	3	6	9	4	2	1	7
7	6	2	3	8	1	4	5	9

DAY 55 TOP

7	5	9	1	3	8	2	4	6
8	4	1	7	6	2	3	9	5
3	6	2	4	9	5	8	7	1
2	8	6	5	4	9	1	3	7
5	3	4	2	7	1	9	6	8
9	1	7	3	8	6	5	2	4
1	7	3	8	2	4	6	5	9
6	2	5	9	1	7	4	8	3
4	9	8	6	5	3	7	1	2

DAY 55 BOTTOM

9	8	5	6	1	3	4	7	2
7	6	1	2	9	4	8	3	5
2	3	4	5	8	7	6	1	9
5	2	8	1	3	6	7	9	4
6	9	3	4	7	5	2	8	1
1	4	7	8	2	9	3	5	6
8	5	6	7	4	1	9	2	3
4	7	9	3	5	2	1	6	8
3	1	2	9	6	8	5	4	7

DAY 56 TOP

9	6	7	8	1	5	2	4	3
5	3	4	9	7	2	8	1	6
1	2	8	6	3	4	5	9	7
4	8	2	3	9	6	7	5	1
6	1	5	2	4	7	9	3	8
3	7	9	5	8	1	6	2	4
7	9	1	4	2	8	3	6	5
2	4	6	7	5	3	1	8	9
8	5	3	1	6	9	4	7	2

DAY 56 BOTTOM

7	3	9	8	6	5	4	2	1
8	5	4	2	3	1	9	7	6
1	2	6	9	7	4	8	5	3
3	4	2	7	8	6	1	9	5
6	7	1	5	4	9	2	3	8
5	9	8	1	2	3	7	6	4
9	1	3	4	5	2	6	8	7
4	6	7	3	9	8	5	1	2
2	8	5	6	1	7	3	4	9

DAY 57 TOP

7	8	1	5	9	4	2	6	3
3	4	6	8	1	2	5	9	7
2	9	5	7	3	6	4	8	1
9	7	4	2	6	1	8	3	5
5	6	2	3	8	7	1	4	9
1	3	8	4	5	9	7	2	6
4	5	3	9	7	8	6	1	2
6	2	7	1	4	3	9	5	8
8	1	9	6	2	5	3	7	4

DAY 57 BOTTOM

8	2	7	5	3	6	4	9	1
1	3	9	2	8	4	5	6	7
5	4	6	9	7	1	2	8	3
7	9	4	8	5	3	1	2	6
3	1	8	6	4	2	9	7	5
6	5	2	1	9	7	8	3	4
2	7	1	4	6	9	3	5	8
9	6	5	3	1	8	7	4	2
4	8	3	7	2	5	6	1	9

DAY 58 TOP

4	1	3	6	8	9	5	2	7
9	2	6	1	5	7	4	8	3
8	7	5	4	2	3	9	6	1
5	6	2	7	1	4	8	3	9
1	9	8	2	3	5	7	4	6
7	3	4	8	9	6	2	1	5
2	5	1	3	7	8	6	9	4
6	8	7	9	4	1	3	5	2
3	4	9	5	6	2	1	7	8

DAY 58 BOTTOM

8	1	3	4	7	5	9	6	2
6	7	2	8	9	1	5	4	3
5	4	9	2	6	3	7	1	8
7	3	4	6	8	9	1	2	5
1	2	8	5	4	7	3	9	6
9	6	5	1	3	2	8	7	4
2	5	7	3	1	4	6	8	9
3	9	6	7	2	8	4	5	1
4	8	1	9	5	6	2	3	7

DAY 59 TOP

8	9	7	4	3	6	1	5	2
3	6	5	2	1	8	4	9	7
4	1	2	9	7	5	3	8	6
1	2	3	5	9	7	6	4	8
9	8	6	3	2	4	7	1	5
7	5	4	8	6	1	9	2	3
2	7	1	6	8	9	5	3	4
5	3	9	7	4	2	8	6	1
6	4	8	1	5	3	2	7	9

DAY 59 BOTTOM

7	9	5	8	6	3	4	2	1
4	8	6	2	1	9	7	3	5
3	1	2	4	7	5	6	9	8
6	7	8	9	2	1	5	4	3
1	4	9	3	5	6	2	8	7
2	5	3	7	8	4	1	6	9
5	3	4	6	9	7	8	1	2
9	2	7	1	4	8	3	5	6
8	6	1	5	3	2	9	7	4

DAY 60 TOP

4	3	5	2	6	1	8	9	7
1	8	9	7	5	3	6	2	4
6	2	7	4	8	9	1	5	3
5	7	1	8	2	4	3	6	9
3	4	2	1	9	6	5	7	8
8	9	6	5	3	7	2	4	1
2	1	4	6	7	8	9	3	5
7	6	3	9	1	5	4	8	2
9	5	8	3	4	2	7	1	6

DAY 60 BOTTOM

4	9	8	6	3	7	2	5	1
2	6	1	8	9	5	7	3	4
7	5	3	4	2	1	8	9	6
8	1	6	5	7	9	4	2	3
9	3	7	2	1	4	6	8	5
5	2	4	3	8	6	9	1	7
3	7	2	1	6	8	5	4	9
6	8	5	9	4	3	1	7	2
1	4	9	7	5	2	3	6	8

DAY **61** TOP

2	5	8	3	4	7	6	9	1
6	1	4	9	5	8	2	7	3
7	3	9	2	6	1	4	8	5
5	9	3	6	8	2	7	1	4
4	8	2	7	1	5	9	3	6
1	7	6	4	9	3	8	5	2
9	2	1	5	7	6	3	4	8
8	6	7	1	3	4	5	2	9
3	4	5	8	2	9	1	6	7

DAY **61** BOTTOM

5	1	3	6	8	4	2	7	9
4	7	9	5	3	2	1	8	6
2	8	6	1	9	7	3	5	4
6	4	1	3	5	8	7	9	2
7	2	5	9	4	6	8	3	1
3	9	8	7	2	1	6	4	5
1	6	4	8	7	9	5	2	3
9	3	7	2	1	5	4	6	8
8	5	2	4	6	3	9	1	7

DAY **62** TOP

3	8	7	5	2	1	6	4	9
2	6	5	8	9	4	1	3	7
4	9	1	7	6	3	8	2	5
1	5	3	2	4	9	7	6	8
9	7	2	1	8	6	4	5	3
8	4	6	3	5	7	2	9	1
7	2	9	4	1	5	3	8	6
6	3	8	9	7	2	5	1	4
5	1	4	6	3	8	9	7	2

DAY **62** BOTTOM

2	5	9	8	4	1	7	3	6
7	3	4	5	6	9	1	2	8
8	1	6	2	3	7	4	9	5
5	4	2	1	8	6	9	7	3
9	8	1	7	5	3	6	4	2
3	6	7	9	2	4	8	5	1
4	7	3	6	1	2	5	8	9
1	2	5	4	9	8	3	6	7
6	9	8	3	7	5	2	1	4

DAY **63** TOP

3	7	9	4	5	2	8	6	1
1	8	6	3	7	9	5	4	2
4	2	5	6	8	1	9	3	7
9	1	2	7	4	5	6	8	3
6	4	3	2	1	8	7	5	9
8	5	7	9	3	6	1	2	4
7	4	3	1	6	4	2	9	5
2	6	4	5	9	7	3	1	8
5	9	1	8	2	3	4	7	6

DAY **63** BOTTOM

7	8	3	4	5	2	9	6	1
2	5	6	1	9	7	8	3	4
4	1	9	6	8	3	5	7	2
6	3	1	9	7	8	4	2	5
9	4	5	2	1	6	7	8	3
8	2	7	5	3	4	6	1	9
1	7	4	8	2	9	3	5	6
5	9	8	3	6	1	2	4	7
3	6	2	7	4	5	1	9	8

DAY 64 TOP

5	6	2	4	1	9	7	8	3
3	1	8	7	6	2	4	9	5
4	7	9	8	5	3	1	6	2
8	3	4	6	9	7	5	2	1
7	5	6	2	8	1	9	3	4
9	2	1	3	4	5	8	7	6
6	4	5	9	3	8	2	1	7
1	8	7	5	2	6	3	4	9
2	9	3	1	7	4	6	5	8

DAY 64 BOTTOM

4	1	3	2	5	8	7	6	9
7	2	9	4	6	3	5	8	1
8	5	6	7	1	9	4	2	3
2	9	8	3	7	4	6	1	5
3	6	5	9	8	1	2	7	4
1	7	4	6	2	5	3	9	8
6	8	1	5	3	2	9	4	7
9	3	2	1	4	7	8	5	6
5	4	7	8	9	6	1	3	2

DAY 65 TOP

9	3	1	8	7	5	2	4	6
5	4	2	1	6	9	7	3	8
8	7	6	4	3	2	1	5	9
4	2	3	5	8	6	9	1	7
7	6	5	9	1	3	8	2	4
1	8	9	2	4	7	3	6	5
6	1	4	7	2	8	5	9	3
2	5	7	3	9	4	6	8	1
3	9	8	6	5	1	4	7	2

DAY 65 BOTTOM

3	4	2	7	5	9	1	6	8
8	9	6	1	4	2	5	7	3
5	7	1	3	6	8	2	9	4
6	8	5	9	3	7	4	2	1
7	2	4	6	1	5	3	8	9
9	1	3	8	2	4	6	5	7
4	5	7	2	9	3	8	1	6
1	3	8	5	7	6	9	4	2
2	6	9	4	8	1	7	3	5

DAY 66 TOP

5	7	8	6	4	1	2	3	9
3	6	4	7	9	2	8	5	1
1	2	9	5	3	8	4	6	7
7	4	1	8	5	9	3	2	6
6	8	3	1	2	7	5	9	4
9	5	2	4	6	3	7	1	8
4	9	6	2	7	5	1	8	3
2	1	7	3	8	6	9	4	5
8	3	5	9	1	4	6	7	2

DAY 66 BOTTOM

6	8	7	3	4	2	1	5	9
2	9	4	1	5	8	7	6	3
1	3	5	9	7	6	2	8	4
4	1	9	7	6	5	8	3	2
7	6	3	8	2	1	9	4	5
5	2	8	4	9	3	6	1	7
3	5	6	2	8	9	4	7	1
8	7	2	5	1	4	3	9	6
9	4	1	6	3	7	5	2	8

DAY 67 TOP

3	9	4	6	8	2	5	1	7
8	7	2	5	4	1	3	9	6
5	1	6	3	9	7	2	8	4
1	2	5	9	6	4	8	7	3
6	3	8	7	2	5	1	4	9
7	4	9	1	3	8	6	2	5
9	5	7	8	1	6	4	3	2
4	8	3	2	5	9	7	6	1
2	6	1	4	7	3	9	5	8

DAY 67 BOTTOM

3	4	8	2	6	5	9	1	7
9	7	2	4	1	8	5	6	3
6	5	1	7	9	3	4	2	8
8	3	4	9	2	1	7	5	6
7	1	6	3	5	4	2	8	9
5	2	9	6	8	7	1	3	4
2	9	3	1	4	6	8	7	5
4	6	5	8	7	2	3	9	1
1	8	7	5	3	9	6	4	2

DAY 68 TOP

2	9	5	3	6	1	8	7	4
6	1	8	5	7	4	2	3	9
4	7	3	2	9	8	1	6	5
8	6	1	4	5	9	3	2	7
3	4	9	6	2	7	5	1	8
7	5	2	8	1	3	4	9	6
1	8	7	9	3	5	6	4	2
9	2	4	1	8	6	7	5	3
5	3	6	7	4	2	9	8	1

DAY 68 BOTTOM

2	3	7	1	5	4	8	9	6
8	6	1	3	9	7	4	5	2
5	4	9	6	8	2	3	1	7
7	2	3	8	1	9	6	4	5
6	9	5	2	4	3	1	7	8
4	1	8	5	7	6	9	2	3
1	8	2	9	3	5	7	6	4
3	5	4	7	6	1	2	8	9
9	7	6	4	2	8	5	3	1

DAY 69 TOP

6	2	7	3	4	5	1	8	9
9	4	8	6	1	7	5	3	2
3	1	5	8	2	9	7	4	6
4	7	2	1	9	8	6	5	3
8	6	1	5	3	2	9	7	4
5	3	9	4	7	6	2	1	8
2	9	4	7	5	3	8	6	1
1	5	6	9	8	4	3	2	7
7	8	3	2	6	1	4	9	5

DAY 69 BOTTOM

9	3	1	6	4	8	2	7	5
7	2	5	3	1	9	6	8	4
8	6	4	7	5	2	3	1	9
4	9	6	1	2	7	8	5	3
3	1	7	5	8	4	9	6	2
2	5	8	9	6	3	1	4	7
1	7	3	4	9	6	5	2	8
6	4	2	8	3	5	7	9	1
5	8	9	2	7	1	4	3	6

DAY 70 TOP

8	7	9	6	2	3	1	5	4
5	2	6	1	4	7	8	9	3
1	3	4	5	8	9	2	7	6
7	1	3	8	6	5	4	2	9
6	4	8	7	9	2	3	1	5
2	9	5	3	1	4	7	6	8
9	5	7	2	3	8	6	4	1
3	6	2	4	5	1	9	8	7
4	8	1	9	7	6	5	3	2

DAY 70 BOTTOM

5	6	9	3	7	6	1	2	8
1	8	3	5	2	9	6	7	4
7	6	2	8	1	4	5	3	9
9	4	5	1	3	2	8	6	7
8	2	7	4	6	5	3	9	1
6	3	1	7	9	8	2	4	5
3	1	4	2	5	7	9	8	6
5	7	6	9	8	3	4	1	2
2	9	8	6	4	1	7	5	3

DAY 71 TOP

1	9	8	7	2	3	5	4	6
4	5	3	1	6	8	7	9	2
2	7	6	4	9	5	1	3	8
9	8	7	6	5	4	3	2	1
3	6	1	9	8	2	4	5	7
5	2	4	3	7	1	8	6	9
6	3	2	8	4	7	9	1	5
8	4	5	2	1	9	6	7	3
7	1	9	5	3	6	2	8	4

DAY 71 BOTTOM

1	3	7	8	6	9	4	2	5
5	8	6	1	4	2	3	9	7
4	9	2	3	7	5	1	8	6
6	7	3	5	8	1	2	4	9
2	5	8	4	9	3	6	7	1
9	1	4	6	2	7	5	3	8
8	6	9	2	1	4	7	5	3
3	4	1	7	5	8	9	6	2
7	2	5	9	3	6	8	1	4

DAY 72 TOP

2	4	8	9	1	7	6	5	3
7	9	6	5	4	3	2	8	1
1	3	5	2	6	8	7	9	4
5	6	7	4	2	9	3	1	8
4	8	1	7	3	6	9	2	5
9	2	3	1	8	5	4	6	7
3	1	2	8	9	4	5	7	6
8	7	4	6	5	2	1	3	9
6	5	9	3	7	1	8	4	2

DAY 72 BOTTOM

2	8	5	9	7	3	1	4	6
3	7	6	1	4	5	9	8	2
1	4	9	8	6	2	3	5	7
9	2	7	6	1	8	4	3	5
6	3	4	5	9	7	8	2	1
8	5	1	2	3	4	6	7	9
4	9	2	7	8	1	5	6	3
7	1	8	3	5	6	2	9	4
5	6	3	4	2	9	7	1	8

DAY **73** TOP

4	3	8	9	1	7	5	2	6
2	5	7	6	8	3	9	1	4
6	1	9	4	2	5	3	7	8
7	6	5	1	4	8	2	9	3
8	4	2	3	7	9	6	5	1
1	9	3	5	6	2	4	8	7
5	7	6	8	9	4	1	3	2
3	2	1	7	5	6	8	4	9
9	8	4	2	3	1	7	6	5

DAY **73** BOTTOM

6	3	5	2	8	7	4	9	1
4	1	9	5	6	3	7	2	8
7	2	8	1	9	4	3	5	6
1	8	3	7	4	9	5	6	2
2	7	6	8	1	5	9	3	4
5	9	4	3	2	6	1	8	7
8	5	2	9	7	1	6	4	3
3	4	1	6	5	8	2	7	9
9	6	7	4	3	2	8	1	5

DAY **74** TOP

4	8	3	2	5	9	1	6	7
5	6	2	7	1	8	9	4	3
1	7	9	6	3	4	5	2	8
9	2	1	5	8	3	4	7	6
7	4	5	1	6	2	3	8	9
6	3	8	4	9	7	2	1	5
2	9	4	3	7	6	8	5	1
3	5	6	8	4	1	7	9	2
8	1	7	9	2	5	6	3	4

DAY **74** BOTTOM

8	6	4	9	7	5	1	3	2
3	5	9	8	2	1	7	4	6
1	2	7	4	3	6	8	9	5
9	1	2	7	5	8	3	6	4
4	3	8	2	6	9	5	1	7
5	7	6	1	4	3	9	2	8
6	9	3	5	8	4	2	7	1
7	8	1	6	9	2	4	5	3
2	4	5	3	1	7	6	8	9

DAY **75** TOP

1	3	8	7	4	9	5	2	6
9	6	7	1	5	2	3	8	4
5	4	2	3	6	8	7	1	9
7	2	9	6	3	4	8	5	1
6	1	5	8	2	7	9	4	3
4	8	3	9	1	5	6	7	2
3	5	6	4	8	1	2	9	7
2	9	4	5	7	3	1	6	8
8	7	1	2	9	6	4	3	5

DAY **75** BOTTOM

4	2	3	6	1	8	7	9	5
1	7	5	4	9	2	6	8	3
9	8	6	5	3	7	1	4	2
5	6	8	2	7	4	9	3	1
7	1	4	9	5	3	8	2	6
2	3	9	1	8	6	4	5	7
3	9	2	7	4	1	5	6	8
8	5	7	3	6	9	2	1	4
6	4	1	8	2	5	3	7	9

DAY **76** TOP

5	7	8	1	4	2	3	9	6
3	2	4	6	5	9	1	7	8
6	1	9	7	8	3	5	4	2
8	3	7	9	6	4	2	5	1
4	5	6	8	2	1	7	3	9
2	9	1	3	7	5	8	6	4
7	8	3	4	1	6	9	2	5
1	4	2	5	9	7	6	8	3
9	6	5	2	3	8	4	1	7

DAY **76** BOTTOM

1	3	5	9	4	7	6	2	8
9	2	7	8	1	6	5	3	4
8	4	6	5	2	3	9	7	1
2	5	3	4	9	8	7	1	6
6	8	1	3	7	5	2	4	9
7	9	4	2	6	1	8	5	3
5	7	9	1	8	4	3	6	2
4	6	8	7	3	2	1	9	5
3	1	2	6	5	9	4	8	7

DAY **77** TOP

3	7	4	8	6	1	5	9	2
8	5	9	2	4	3	7	6	1
2	1	6	5	7	9	3	8	4
4	9	5	7	1	5	2	3	6
5	2	3	6	9	4	1	7	8
1	6	7	3	8	2	4	5	9
6	8	1	4	3	7	9	2	5
7	4	2	9	5	6	8	1	3
9	3	5	1	2	8	6	4	7

DAY **77** BOTTOM

5	3	9	6	2	1	8	4	7
1	7	2	9	4	8	3	6	5
8	4	6	7	5	3	2	9	1
4	1	5	3	6	2	7	8	9
6	8	7	5	1	9	4	3	2
9	2	3	8	7	4	5	1	6
2	9	4	1	3	7	6	5	8
3	6	8	2	9	5	1	7	4
7	5	1	4	8	6	9	2	3

DAY **78** TOP

6	5	3	7	4	1	9	2	8
1	4	9	2	8	3	5	6	7
7	8	2	5	6	9	1	3	4
2	9	5	6	1	4	8	7	3
8	3	1	9	7	2	4	5	6
4	6	7	3	5	8	2	9	1
5	7	8	1	9	6	3	4	2
9	2	4	8	3	7	6	1	5
3	1	6	4	2	5	7	8	9

DAY **78** BOTTOM

5	6	7	1	9	4	2	3	8
1	3	8	5	2	6	7	9	4
9	2	4	7	8	3	6	1	5
6	1	3	8	5	7	9	4	2
7	8	9	3	4	2	1	5	6
2	4	5	9	6	1	3	8	7
8	5	1	6	7	9	4	2	3
3	7	2	4	1	5	8	6	9
4	9	6	2	3	8	5	7	1

DAY 79 TOP

2	4	9	1	6	3	5	8	7
3	8	6	7	5	2	4	1	9
5	1	7	8	4	9	2	6	3
6	5	8	9	2	1	3	7	4
9	7	3	6	8	4	1	5	2
1	2	4	5	3	7	6	9	8
8	3	5	4	7	6	9	2	1
4	6	1	2	9	8	7	3	5
7	9	2	3	1	5	8	4	6

DAY 79 BOTTOM

2	5	6	1	4	8	7	3	9
1	3	4	2	9	7	5	8	6
7	8	9	5	3	6	4	2	1
9	7	1	8	5	4	2	6	3
4	2	5	3	6	9	1	7	8
3	6	8	7	1	2	9	5	4
6	1	3	9	2	5	8	4	7
5	9	7	4	8	3	6	1	2
8	4	2	6	7	1	3	9	5

DAY 80 TOP

2	4	8	9	1	7	6	5	3
7	9	6	5	4	3	2	8	1
1	3	5	2	6	8	7	9	4
5	6	7	4	2	9	3	1	8
4	8	1	7	3	6	9	2	5
9	2	3	1	8	5	4	6	7
3	1	2	8	9	4	5	7	6
8	7	4	6	5	2	1	3	9
6	5	9	3	7	1	8	4	2

DAY 80 BOTTOM

2	5	4	9	6	3	8	7	1
1	8	6	5	2	7	9	3	4
7	9	3	4	1	8	2	6	5
4	3	5	7	8	9	6	1	2
9	6	2	1	4	5	7	8	3
8	7	1	2	3	6	4	5	9
3	2	7	6	9	1	5	4	8
6	1	9	8	5	4	3	2	7
5	4	8	3	7	2	1	9	6

DAY 81 TOP

7	6	2	3	4	9	1	5	3
4	9	5	8	1	2	3	6	7
1	8	3	7	5	6	9	2	4
2	3	6	4	8	7	5	9	1
8	5	7	2	9	1	6	4	3
9	4	1	5	6	3	7	8	2
3	2	4	6	7	5	8	1	9
5	7	9	1	2	8	4	3	6
6	1	8	9	3	4	2	7	5

DAY 81 BOTTOM

2	1	4	7	8	5	9	6	3
6	9	3	4	2	1	7	8	5
8	5	7	3	6	9	2	4	1
7	6	1	5	3	8	4	9	2
5	4	2	9	1	7	8	3	6
3	8	9	6	4	2	1	5	7
4	3	8	2	7	6	5	1	9
1	7	5	8	9	3	6	2	4
9	2	6	1	5	4	3	7	8

DAY **82** TOP

5	3	4	7	8	6	9	1	2
2	8	7	1	4	9	6	3	5
1	6	9	5	3	2	8	7	4
8	1	2	3	6	4	7	5	9
4	5	3	9	1	7	2	8	6
9	7	6	8	2	5	3	4	1
6	9	5	4	7	8	1	2	3
7	4	1	2	9	3	5	6	8
3	2	8	6	5	1	4	9	7

DAY **82** BOTTOM

1	7	6	8	3	9	5	4	2
8	5	3	2	1	4	9	6	7
9	2	4	5	7	6	3	1	8
7	4	9	6	5	2	8	3	1
2	6	1	3	9	8	7	5	4
3	8	5	7	4	1	2	9	6
6	1	2	9	8	3	4	7	5
5	9	8	4	6	7	1	2	3
4	3	7	1	2	5	6	8	9

DAY **83** TOP

4	2	6	8	5	1	9	7	3
9	3	1	2	6	7	4	5	8
8	5	7	4	9	3	2	1	6
6	1	5	7	4	9	8	3	2
2	9	4	3	1	8	7	6	5
7	8	3	6	2	5	1	9	4
5	4	2	1	7	8	3	8	9
1	6	8	9	3	2	5	4	7
3	7	9	5	8	4	6	2	1

DAY **83** BOTTOM

6	2	7	5	9	1	8	4	3
5	1	3	4	2	8	9	7	6
8	9	4	3	6	7	1	2	5
2	8	9	1	5	4	3	6	7
4	5	6	9	7	3	2	8	1
7	3	1	2	8	6	5	9	4
9	4	2	7	1	5	6	3	8
3	6	5	8	4	2	7	1	9
1	7	8	6	3	9	4	5	3

DAY **84** TOP

3	2	4	6	9	8	5	1	7
1	5	9	2	3	7	8	4	6
7	6	8	5	4	1	2	9	3
5	3	6	8	1	9	7	2	4
4	1	7	3	5	2	6	8	9
8	9	2	4	7	6	3	5	1
9	4	3	7	2	5	1	6	7
6	7	5	1	8	4	9	3	2
2	8	1	9	6	3	4	7	5

DAY **84** BOTTOM

5	1	3	9	6	4	8	2	7
7	2	6	1	5	8	9	4	3
9	4	8	2	3	7	6	1	5
4	6	7	5	9	3	2	8	1
2	3	5	4	8	1	7	6	9
1	8	9	7	2	6	5	3	4
8	5	4	6	1	9	3	7	2
6	7	2	3	4	5	1	9	8
3	9	1	8	7	2	4	5	6

DAY **85** TOP

6	3	2	8	5	9	4	1	7
1	8	4	7	6	3	5	2	9
9	5	7	2	1	4	6	3	8
8	7	6	9	4	2	1	5	3
2	1	9	3	8	5	7	4	6
5	4	3	1	7	6	8	9	2
7	2	8	4	3	1	9	6	5
4	9	5	6	2	8	3	7	1
3	6	1	5	9	7	2	8	4

DAY **85** BOTTOM

5	2	1	8	4	3	6	9	7
3	8	9	6	1	7	4	5	2
6	4	7	9	2	5	1	8	3
9	7	8	2	5	1	3	4	6
1	3	4	7	6	8	5	2	9
2	5	6	3	9	4	7	1	8
8	9	5	1	7	6	2	3	4
7	1	2	4	3	9	8	6	5
4	6	3	5	8	2	9	7	1

DAY **86** TOP

3	5	9	1	2	8	7	6	4
8	1	7	6	5	4	3	9	2
2	4	6	3	7	9	8	1	5
6	7	8	5	3	1	4	2	9
5	9	2	8	4	7	1	3	6
1	3	4	2	9	6	5	7	8
4	2	3	9	1	5	6	8	7
9	8	5	7	6	3	2	4	1
7	6	1	4	8	2	9	5	3

DAY **86** BOTTOM

3	9	6	1	8	4	2	5	7
4	8	7	2	5	6	1	9	3
2	5	1	9	7	3	4	6	8
1	3	8	7	2	9	5	4	6
7	4	5	6	1	8	9	3	2
9	6	2	3	4	5	7	8	1
5	1	3	8	9	2	6	7	4
8	2	9	4	6	7	3	1	5
6	7	4	5	3	1	8	2	9

DAY **87** TOP

1	3	7	8	9	6	5	4	2
6	8	5	4	3	2	1	7	9
9	2	4	1	5	7	6	8	3
4	5	6	3	1	8	2	9	7
3	7	9	6	2	5	8	1	4
8	1	2	9	7	4	3	5	6
2	9	1	7	8	3	4	6	5
7	6	3	5	4	1	9	2	8
5	4	8	2	6	9	7	3	1

DAY **87** BOTTOM

4	1	7	2	9	5	3	6	8
5	9	8	3	6	7	2	1	4
3	6	2	1	8	4	5	7	9
2	4	9	8	3	1	6	5	7
8	5	6	7	2	9	1	4	3
1	7	3	4	5	6	8	9	2
6	2	4	9	1	3	7	8	5
9	3	1	5	7	8	4	2	6
7	8	5	6	4	2	9	3	1

DAY 88 TOP

2	4	8	9	1	7	6	5	3
7	9	6	5	4	3	2	8	1
1	3	5	2	6	8	7	9	4
5	6	7	4	2	9	3	1	8
4	8	1	7	3	6	9	2	5
9	2	3	1	8	5	4	6	7
3	1	2	8	9	4	5	7	6
8	7	4	6	5	2	1	3	9
6	5	9	3	7	1	8	4	2

DAY 88 BOTTOM

5	2	8	3	1	6	4	7	9
6	1	9	4	7	8	3	2	5
4	7	3	2	9	5	6	8	1
3	5	1	9	4	2	7	6	8
9	6	7	8	3	1	2	5	4
2	8	4	5	6	7	9	1	3
7	3	5	1	2	4	8	9	6
1	4	2	6	8	9	5	3	7
8	9	6	7	5	3	1	4	2

DAY 89 TOP

8	6	2	1	9	3	4	5	7
3	1	4	5	6	7	8	2	9
9	7	5	8	4	2	3	1	6
5	4	3	6	8	1	7	9	2
6	2	9	3	7	4	1	8	5
1	8	7	9	2	5	6	4	3
7	9	8	2	1	6	5	3	4
2	3	6	4	5	8	9	7	1
4	5	1	7	3	9	2	6	8

DAY 89 BOTTOM

5	8	2	7	9	4	6	3	1
4	9	1	6	3	2	7	8	5
6	3	7	8	1	5	4	2	9
7	5	9	1	6	8	3	4	2
1	4	3	2	7	9	8	5	6
8	2	6	5	4	3	1	9	7
3	7	5	9	8	6	2	1	4
9	6	8	4	2	1	5	7	3
2	1	4	3	5	7	9	6	8

DAY 90 TOP

9	7	3	2	1	4	5	6	8
4	2	5	6	7	8	9	3	1
1	8	6	9	5	3	4	2	7
6	5	4	7	9	2	8	1	3
7	3	1	4	8	5	2	9	6
2	9	8	1	3	6	7	5	4
8	1	9	3	2	7	6	4	5
3	4	7	5	6	9	1	8	2
5	6	2	8	4	1	3	7	9

DAY 90 BOTTOM

6	9	3	8	1	5	7	4	2
5	1	2	7	4	3	8	9	6
7	4	8	9	2	6	5	3	1
8	6	1	2	7	9	4	5	3
2	5	4	3	8	1	9	6	7
9	3	7	6	5	4	2	1	8
4	8	6	1	9	7	3	2	5
1	7	9	5	3	2	6	8	4
3	2	5	4	6	8	1	7	9

DAY **91** TOP

5	4	9	7	2	6	1	8	3
1	2	3	4	8	9	5	7	6
8	6	7	5	3	1	9	2	4
7	3	8	1	5	4	6	9	2
6	5	2	8	9	3	7	4	1
9	1	4	2	6	7	3	5	8
2	7	1	6	4	5	8	3	9
3	8	5	9	1	2	4	6	7
4	9	6	3	7	8	2	1	5

DAY **91** BOTTOM

9	6	7	8	3	2	4	5	1
1	8	5	7	4	9	6	2	3
4	2	3	1	6	5	7	8	9
3	7	6	2	8	1	5	9	4
2	5	1	4	9	7	8	3	6
8	9	4	6	5	3	1	7	2
5	1	9	3	7	6	2	4	8
6	3	8	5	2	4	9	1	7
7	4	2	9	1	8	3	6	5

DAY **92** TOP

9	3	7	5	6	8	4	2	1
1	5	6	9	2	4	7	3	8
2	4	8	7	3	1	6	9	5
5	1	4	2	7	9	8	6	3
6	8	9	1	4	3	2	5	7
3	7	2	8	5	6	1	4	9
8	2	5	6	9	7	3	1	4
4	9	1	3	8	2	5	7	6
7	6	3	4	1	5	9	8	2

DAY **92** BOTTOM

7	6	3	1	9	8	5	2	4
2	4	8	5	3	7	9	6	1
9	1	5	2	4	6	3	8	7
8	7	2	3	1	5	4	9	6
6	5	4	8	7	9	2	1	3
1	3	9	6	2	4	7	5	8
3	2	6	4	5	1	8	7	9
4	9	1	7	8	2	6	3	5
5	8	7	9	6	3	1	4	2

DAY **93** TOP

3	8	6	4	7	2	1	5	9
7	1	2	5	9	3	6	8	4
4	5	9	6	8	1	3	2	7
8	4	1	3	2	9	5	7	6
9	7	5	1	6	8	4	3	2
6	2	3	7	5	4	9	1	8
5	9	8	2	3	6	7	4	1
2	3	4	9	1	7	8	6	5
1	6	7	8	4	5	2	9	3

DAY **93** BOTTOM

8	4	9	5	1	6	7	2	3
1	7	6	3	2	8	9	5	4
5	3	2	9	4	7	8	1	6
2	1	3	7	9	4	5	6	8
4	5	7	8	6	2	3	9	1
9	6	8	1	3	5	2	4	7
7	9	1	4	5	3	6	8	2
6	8	5	2	7	1	4	3	9
3	2	4	6	8	9	1	7	5

DAY 94 TOP

9	8	1	5	3	7	2	4	6
4	7	2	8	6	9	3	1	5
5	6	3	2	4	1	9	7	8
1	2	6	9	7	5	8	3	4
3	5	8	4	1	2	7	6	9
7	4	9	3	8	6	1	5	2
2	3	5	7	9	4	6	8	1
8	1	4	6	2	3	5	9	7
6	9	7	1	5	8	4	2	3

DAY 94 BOTTOM

8	3	9	2	4	7	6	1	5
6	7	1	8	3	5	2	9	4
4	2	5	1	9	6	3	8	7
7	9	2	6	5	8	1	4	3
5	6	4	3	1	9	7	2	8
3	1	8	4	7	2	5	6	9
2	8	3	5	6	4	9	7	1
9	5	6	7	8	1	4	3	2
1	4	7	9	2	3	8	5	6

DAY 95 TOP

8	9	2	7	4	6	5	1	3
3	4	1	9	8	5	7	6	2
7	5	6	1	2	3	9	8	4
4	3	8	6	5	9	1	2	7
2	1	5	8	7	4	3	9	6
6	7	9	3	1	2	8	4	5
9	2	3	5	6	1	4	7	8
1	6	7	4	3	8	2	5	9
5	8	4	2	9	7	6	3	1

DAY 95 BOTTOM

7	9	8	6	5	4	3	1	2
3	4	2	9	1	7	8	6	5
6	1	5	2	3	8	4	9	7
4	3	7	5	9	1	6	2	8
5	6	9	3	8	2	1	7	4
8	2	1	7	4	6	5	3	9
2	5	6	4	7	3	9	8	1
9	8	3	1	2	5	7	4	6
1	7	4	8	6	9	2	5	3

DAY 96 TOP

6	8	9	4	5	2	7	1	3
7	4	3	1	6	8	5	2	9
2	5	1	7	3	9	8	4	6
1	7	2	9	8	3	4	6	5
4	9	5	6	7	1	3	8	2
8	3	6	5	2	4	1	9	7
9	1	7	3	4	6	2	5	8
5	6	8	2	1	7	9	3	4
3	2	4	8	9	5	6	7	1

DAY 96 BOTTOM

8	7	1	9	5	2	6	4	3
3	4	5	6	7	1	2	8	9
9	2	6	4	8	3	1	5	7
7	5	4	1	6	9	3	2	8
6	8	3	5	2	7	4	9	1
1	9	2	8	3	4	7	6	5
2	3	9	7	4	5	8	1	6
5	6	7	2	1	8	9	3	4
4	1	8	3	9	6	5	7	2

DAY **97** TOP

3	9	2	7	1	6	5	8	4
4	6	5	2	9	8	7	1	3
8	7	1	4	3	5	9	6	2
1	5	8	9	4	3	6	2	7
2	4	9	1	6	7	8	3	5
7	3	6	5	8	2	4	9	1
9	2	7	8	5	1	3	4	6
5	8	3	6	2	4	1	7	9
6	1	4	3	7	9	2	5	8

DAY **97** BOTTOM

9	4	7	5	8	1	2	3	6
6	3	1	2	7	9	8	5	4
8	5	2	6	3	4	7	9	1
1	9	6	7	5	2	3	4	8
5	7	8	4	6	3	9	1	2
4	2	3	1	9	8	6	7	5
7	6	4	3	2	5	1	8	9
3	8	5	9	1	6	4	2	7
2	1	9	8	4	7	5	6	3

DAY **98** TOP

7	2	4	1	5	6	9	3	8
6	3	1	2	9	8	7	5	4
5	8	9	7	3	4	6	1	2
2	6	5	3	4	9	8	7	1
8	4	7	5	6	1	3	2	9
9	1	3	8	2	7	5	4	6
4	7	6	9	1	3	2	8	5
3	9	2	4	8	5	1	6	7
1	5	8	6	7	2	4	9	3

DAY **98** BOTTOM

3	2	9	4	5	8	6	7	1
1	8	7	3	2	6	5	9	4
5	4	6	7	9	1	8	3	2
6	1	5	2	7	9	3	4	8
2	7	3	5	8	4	9	1	6
8	9	4	1	6	3	2	5	7
4	3	2	6	1	5	7	8	9
9	6	1	8	3	7	4	2	5
7	5	8	9	4	2	1	6	3

DAY **99** TOP

9	7	6	5	4	8	3	2	1
2	4	5	1	9	3	6	8	7
1	3	8	7	6	2	5	4	9
7	6	1	4	8	5	2	9	3
4	2	3	6	1	9	8	7	5
5	8	9	3	2	7	1	6	4
8	1	4	9	5	6	7	3	2
3	9	2	8	7	1	4	5	6
6	5	7	2	3	4	9	1	8

DAY **99** BOTTOM

8	6	9	5	3	4	7	1	2
7	4	3	1	8	2	5	6	9
2	5	1	7	6	9	3	4	8
9	2	6	4	5	3	8	7	1
3	1	7	8	2	6	9	5	4
4	8	5	9	7	1	6	2	3
6	9	4	3	1	7	2	8	5
1	7	8	2	9	5	4	3	6
5	3	2	6	4	8	1	9	7

DAY **100** TOP

8	4	3	1	6	5	7	9	2
2	6	9	4	8	7	1	5	3
1	7	5	9	3	2	4	8	6
3	9	7	8	1	6	2	4	5
6	2	8	5	7	4	9	3	1
5	1	4	2	9	3	8	6	7
7	8	6	3	4	1	5	2	9
4	3	2	7	5	9	6	1	8
9	5	1	6	2	8	3	7	4

DAY **100** BOTTOM

7	8	4	6	9	2	1	5	3
1	2	6	3	4	5	7	8	9
9	3	5	1	8	7	2	6	4
3	9	8	7	1	6	5	4	2
2	6	7	4	5	3	8	9	1
5	4	1	9	2	8	3	7	6
8	1	9	2	7	4	6	3	5
6	5	2	8	3	9	4	1	7
4	7	3	5	6	1	9	2	8

DAY **101** TOP

9	1	8	7	4	5	6	2	3
6	2	5	9	8	3	7	1	4
4	3	7	2	1	6	8	9	5
8	9	2	4	5	7	3	6	1
5	7	1	6	3	9	2	4	8
3	4	6	8	2	1	5	7	9
7	8	9	3	6	4	1	5	2
2	5	4	1	7	8	9	3	6
1	6	3	5	9	2	4	8	7

DAY **101** BOTTOM

1	7	4	6	5	2	3	9	8
3	8	2	9	4	7	6	5	1
6	5	9	8	3	1	7	4	2
4	6	5	1	8	3	2	7	9
9	1	3	2	7	5	4	8	6
7	2	8	4	6	9	1	3	5
2	9	7	5	1	4	8	6	3
8	4	1	3	9	6	5	2	7
5	3	6	7	2	8	9	1	4

DAY **102** TOP

1	6	9	5	3	4	8	2	7
7	5	3	8	6	2	4	1	9
2	4	8	1	7	9	5	3	6
4	1	7	3	8	6	9	5	2
5	8	2	9	1	7	3	6	4
3	9	6	2	4	5	1	7	8
8	2	1	6	9	3	7	4	5
6	3	4	7	5	8	2	9	1
9	7	5	4	2	1	6	8	3

DAY **102** BOTTOM

1	8	5	9	7	3	6	4	2
6	4	9	1	5	2	8	3	7
7	3	2	8	4	6	9	1	5
9	1	6	2	3	5	7	8	4
8	2	4	7	1	9	3	5	6
3	5	7	4	6	8	2	9	1
2	9	1	5	8	7	4	6	3
5	6	8	3	2	4	1	7	9
4	7	3	6	9	1	5	2	8

DAY **103** TOP

9	3	4	6	2	1	8	7	5
5	8	1	7	4	9	6	2	3
2	6	7	5	8	3	9	4	1
3	7	8	1	9	2	4	5	6
1	9	5	4	6	7	3	8	2
6	4	2	3	5	8	1	9	7
4	5	3	8	7	6	2	1	9
8	2	6	9	1	5	7	3	4
7	1	9	2	3	4	5	6	8

DAY **103** BOTTOM

3	4	8	9	6	2	1	7	5
7	5	6	4	3	1	8	9	2
2	1	9	7	5	8	3	6	4
5	9	7	3	1	4	6	2	8
1	6	4	2	8	7	5	3	9
8	2	3	5	9	6	7	4	1
6	8	2	1	7	9	4	5	3
9	3	1	6	4	5	2	8	7
4	7	5	8	2	3	9	1	6

DAY **104** TOP

5	3	9	8	6	4	7	1	2
1	4	7	2	3	5	8	9	6
6	2	8	1	7	9	5	4	3
8	6	3	9	1	7	4	2	5
4	1	5	3	8	2	9	6	7
9	7	2	4	5	6	3	8	1
2	5	4	6	9	3	1	7	8
7	8	6	5	4	1	2	3	9
3	9	1	7	2	8	6	5	4

DAY **104** BOTTOM

5	1	8	7	3	9	4	2	6
2	7	4	6	1	8	9	5	3
6	9	3	4	5	2	1	8	7
3	4	6	5	2	1	7	9	8
9	5	7	8	6	4	2	3	1
8	2	1	9	7	3	6	4	5
4	6	5	2	8	7	3	1	9
1	8	9	3	4	6	5	7	2
7	3	2	1	9	5	8	6	4

DAY **105** TOP

4	9	5	3	6	2	7	8	1
7	2	6	8	1	5	9	4	3
8	1	3	9	4	7	5	2	6
2	5	9	6	8	3	1	7	4
3	7	4	5	2	1	8	6	9
6	8	1	7	9	4	3	5	2
5	3	2	1	7	6	4	9	8
9	6	7	4	3	8	2	1	5
1	4	8	2	5	9	6	3	7

DAY **105** BOTTOM

3	7	4	5	9	2	1	8	6
1	9	8	6	4	7	5	3	2
5	6	2	8	3	1	4	9	7
7	2	5	3	1	8	9	6	4
4	8	9	7	6	5	3	2	1
6	1	3	9	2	4	7	5	8
8	3	7	1	5	6	2	4	9
2	5	1	4	8	9	6	7	3
9	4	6	2	7	3	8	1	5

DAY 106 TOP

3	5	4	9	6	1	2	7	8
2	1	6	5	7	8	3	4	9
9	7	8	4	3	2	6	1	5
6	3	1	7	8	5	4	9	2
8	4	9	2	1	3	7	5	6
7	2	5	6	4	9	8	3	1
4	9	2	3	5	6	1	8	7
5	8	3	1	2	7	9	6	4
1	6	7	8	9	4	5	2	3

DAY 106 BOTTOM

1	3	7	2	4	5	6	9	8
9	5	8	6	7	1	3	4	2
2	4	6	3	9	8	1	5	7
8	9	4	7	1	6	5	2	3
3	1	2	5	8	4	9	7	6
6	7	5	9	2	3	8	1	4
4	6	1	8	5	7	2	3	9
7	8	9	1	3	2	4	6	5
5	2	3	4	6	9	7	8	1

DAY 107 TOP

5	3	4	1	7	8	2	9	6
6	8	1	5	2	9	3	4	7
9	7	2	3	6	4	8	5	1
8	5	7	6	4	2	9	1	3
1	2	6	9	5	3	7	8	4
4	9	3	7	8	1	6	2	5
7	1	9	8	3	5	4	6	2
3	4	8	2	1	6	5	7	9
2	6	5	4	9	7	1	3	8

DAY 107 BOTTOM

2	9	8	5	6	4	1	7	3
6	5	1	2	7	3	9	4	8
4	7	3	1	8	9	2	6	5
8	1	4	7	5	2	3	9	6
7	3	6	8	9	1	4	5	2
9	2	5	4	3	6	7	8	1
3	8	7	9	2	5	6	1	4
1	6	9	3	4	8	5	2	7
5	4	2	6	1	7	8	3	9

DAY 108 TOP

9	4	3	7	1	2	6	8	5
2	7	6	5	8	4	9	3	1
5	1	8	3	9	6	4	2	7
8	5	7	9	6	3	1	4	2
4	9	2	8	7	1	5	6	3
6	3	1	2	4	5	8	7	9
3	8	5	4	2	9	7	1	6
1	2	4	6	5	7	3	9	8
7	6	9	1	3	8	2	5	4

DAY 108 BOTTOM

4	9	8	3	5	7	2	6	1
1	3	6	8	4	2	9	5	7
2	5	7	1	6	9	8	3	4
3	4	2	9	7	6	1	8	5
7	8	9	5	1	3	6	4	2
5	6	1	2	8	4	3	7	9
8	2	5	4	3	1	7	9	6
9	7	3	6	2	5	4	1	8
6	1	4	7	9	8	5	2	3

DAY **109** TOP

7	2	1	8	4	9	5	6	3
4	8	5	6	2	3	7	9	1
6	9	3	1	5	7	2	8	4
2	7	4	3	9	8	6	1	5
5	1	9	4	6	2	3	7	8
3	6	8	7	1	5	9	4	2
1	3	6	2	7	4	8	5	9
8	5	7	9	3	1	4	2	6
9	4	2	5	8	6	1	3	7

DAY **109** BOTTOM

7	2	4	5	9	3	6	8	1
1	3	5	7	6	8	2	4	9
8	9	6	2	1	4	3	7	5
5	6	1	8	7	2	9	3	4
3	7	9	1	4	6	8	5	2
4	8	2	9	3	5	1	6	7
2	4	3	6	5	1	7	9	8
6	1	7	4	8	9	5	2	3
9	5	8	3	2	7	4	1	6

DAY **110** TOP

9	1	5	4	7	2	8	6	3
7	8	2	9	3	6	1	5	4
6	4	3	5	8	1	2	7	9
5	3	4	8	1	7	6	9	2
1	2	7	6	9	3	5	4	8
8	6	9	2	4	5	3	1	7
2	7	6	3	5	4	9	8	1
4	5	8	1	2	9	7	3	6
3	9	1	7	6	8	4	2	5

DAY **110** BOTTOM

2	8	1	5	3	9	7	6	4
7	4	9	1	2	6	3	5	8
5	6	3	8	4	7	1	9	2
9	3	8	6	7	2	4	1	5
6	5	7	4	1	8	9	2	3
1	2	4	9	5	3	6	8	7
3	1	2	7	6	5	8	4	9
8	7	6	2	9	4	5	3	1
4	9	5	3	8	1	2	7	6

DAY **111** TOP

6	2	3	4	5	8	1	7	9
1	7	8	6	9	2	5	4	3
9	5	4	7	3	1	6	8	2
4	1	7	2	6	9	3	5	8
8	9	6	5	7	3	4	2	1
5	3	2	1	8	4	9	6	7
2	4	9	8	1	6	7	3	5
7	6	1	3	2	5	8	9	4
3	8	5	9	4	7	2	1	6

DAY **111** BOTTOM

3	7	1	4	2	5	6	8	9
5	9	8	3	1	6	4	2	7
2	6	4	7	9	8	5	3	1
7	2	5	6	8	1	3	9	4
9	1	3	5	4	2	7	6	8
8	4	6	9	3	7	1	5	2
6	8	7	1	5	9	2	4	3
4	5	9	2	7	3	8	1	6
1	3	2	8	6	4	9	7	5

DAY **112** TOP

3	5	4	1	2	8	6	9	7
8	1	6	9	5	7	4	2	3
9	7	2	6	3	4	8	5	1
1	4	8	5	7	9	3	6	2
2	3	9	4	6	1	7	8	5
5	6	7	2	8	3	9	1	4
6	2	3	8	4	5	1	7	9
4	9	5	7	1	6	2	3	8
7	8	1	3	9	2	5	4	6

DAY **112** BOTTOM

1	9	5	8	6	7	2	3	4
2	4	8	9	1	3	7	5	6
3	7	6	2	4	5	1	9	8
8	3	1	5	7	4	6	2	9
6	2	4	3	9	1	5	8	7
7	5	9	6	2	8	3	4	1
4	1	2	7	5	9	8	6	3
9	6	3	1	8	2	4	7	5
5	8	7	4	3	6	9	1	2

DAY **113** TOP

1	2	4	5	9	7	3	8	6
3	9	6	2	8	1	7	5	4
5	7	8	4	3	6	2	9	1
8	3	5	6	4	2	9	1	7
2	6	1	9	7	5	8	4	3
9	4	7	8	1	3	6	2	5
6	5	9	7	2	4	1	3	8
4	8	3	1	6	9	5	7	2
7	1	2	3	5	8	4	6	9

DAY **113** BOTTOM

7	8	3	9	1	6	5	2	4
4	6	9	8	2	5	1	7	3
2	1	5	4	3	7	6	9	8
6	5	2	7	4	3	9	8	1
9	3	8	1	6	2	7	4	5
1	7	4	5	8	9	3	6	2
3	4	6	2	7	1	8	5	9
8	9	1	6	5	4	2	3	7
5	2	7	3	9	8	4	1	6

DAY **114** TOP

2	4	7	8	6	3	5	9	1
1	3	6	9	2	5	7	4	8
5	8	9	1	4	7	2	3	6
6	9	1	2	5	4	3	8	7
3	2	8	7	1	9	6	5	4
7	5	4	3	8	6	9	1	2
4	7	3	6	9	1	8	2	5
9	1	2	5	7	8	4	6	3
8	6	5	4	3	2	1	7	9

DAY **114** BOTTOM

5	3	7	1	6	4	8	2	9
9	1	8	2	3	5	7	6	4
6	4	2	9	7	8	5	1	3
1	7	4	5	2	9	6	3	8
2	9	6	7	8	3	4	5	1
3	8	5	4	1	6	9	7	2
8	6	1	3	4	7	2	9	5
7	2	9	8	5	1	3	4	6
4	5	3	6	9	2	1	8	7

DAY **115** TOP

1	2	3	5	6	7	9	8	4
9	7	4	3	2	8	1	6	5
8	5	6	9	1	4	3	7	2
6	4	8	2	9	1	5	3	7
5	1	7	4	8	3	6	2	9
3	9	2	6	7	5	4	1	8
4	8	1	7	3	9	2	5	6
2	3	5	8	4	6	7	9	1
7	6	9	1	5	2	8	4	3

DAY **115** BOTTOM

6	1	4	3	2	9	5	8	7
3	2	8	5	7	1	6	4	9
7	5	9	4	8	6	2	3	1
2	4	3	7	1	5	9	6	8
9	6	7	2	4	8	1	5	3
1	8	5	6	9	3	4	7	2
5	3	2	9	6	7	8	1	4
8	9	6	1	3	4	7	2	5
4	7	1	8	5	2	3	9	6

DAY **116** TOP

1	8	4	3	2	5	6	7	9
5	3	6	7	8	9	1	4	2
2	9	7	1	6	4	5	3	8
7	6	5	8	1	3	9	2	4
8	4	2	5	9	6	3	1	7
3	1	9	2	4	7	8	6	5
9	2	1	4	3	8	7	5	6
4	5	8	6	7	1	2	9	3
6	7	3	9	5	2	4	8	1

DAY **116** BOTTOM

7	1	4	9	2	6	8	5	3
6	2	3	8	5	4	9	1	7
8	5	9	1	3	7	6	4	2
9	7	2	3	8	1	5	6	4
3	6	5	4	9	2	1	7	8
1	4	8	7	6	5	3	2	9
5	9	7	2	1	8	4	3	6
2	8	1	6	4	3	7	9	5
4	3	6	5	7	9	2	8	1

DAY **117** TOP

2	9	5	4	3	6	7	8	1
6	4	7	8	9	1	2	5	3
3	1	8	2	7	5	6	4	9
8	7	6	9	2	4	1	3	5
9	5	3	6	1	7	4	2	8
4	2	1	3	5	8	9	7	6
1	3	2	5	4	9	8	6	7
5	6	9	7	8	2	3	1	4
7	8	4	1	6	3	5	9	2

DAY **117** BOTTOM

8	2	5	1	3	7	9	6	4
7	3	4	9	6	5	1	2	8
9	6	1	2	4	8	7	5	3
1	8	3	4	9	2	6	7	5
4	7	6	5	1	3	2	8	9
2	5	9	8	7	6	4	3	1
6	1	8	3	2	9	5	4	7
3	9	2	7	5	4	8	1	6
5	4	7	6	8	1	3	9	2

DAY 118 TOP

3	1	6	5	4	7	8	9	2
7	5	8	9	1	2	3	6	4
4	2	9	3	8	6	7	5	1
9	8	7	1	3	5	2	4	6
1	6	4	7	2	8	5	3	9
5	3	2	4	6	9	1	8	7
2	4	3	6	5	1	9	7	8
6	7	1	8	9	3	4	2	5
8	9	5	2	7	4	6	1	3

DAY 118 BOTTOM

9	3	6	2	4	8	1	7	5
8	4	5	1	7	6	2	3	9
1	7	2	3	5	9	8	6	4
2	9	4	5	1	3	7	8	6
5	8	7	6	2	4	3	9	1
3	6	1	9	8	7	5	4	2
7	2	9	4	3	1	6	5	8
4	1	3	8	6	5	9	2	7
6	5	8	7	9	2	4	1	3

DAY 119 TOP

4	2	7	6	5	8	9	1	3
8	6	9	1	2	3	4	7	5
5	3	1	4	9	7	8	6	2
1	9	8	2	4	6	3	5	7
2	7	5	8	3	9	6	4	1
6	4	3	5	7	1	2	9	8
3	5	4	7	6	2	1	8	9
7	8	2	9	1	4	5	3	6
9	1	6	3	8	5	7	2	4

DAY 119 BOTTOM

1	7	4	8	6	2	9	3	5
2	6	5	9	3	4	8	7	1
9	3	8	7	5	1	2	4	6
8	1	6	5	9	7	3	2	4
5	2	3	4	8	6	7	1	9
7	4	9	1	2	3	5	6	8
3	8	1	6	7	9	4	5	2
6	9	7	2	4	5	1	8	3
4	5	2	3	1	8	6	9	7

DAY 120 TOP

5	3	8	7	6	9	1	2	4
9	8	1	2	3	4	5	8	6
6	4	2	5	1	8	9	7	3
2	1	9	3	5	7	4	6	8
3	8	6	9	4	1	7	5	2
7	5	4	6	8	2	3	1	9
4	6	5	8	7	3	2	9	1
8	9	3	1	2	5	6	4	7
1	2	7	4	9	6	8	3	5

DAY 120 BOTTOM

2	8	5	9	7	3	1	4	6
3	7	6	1	4	5	9	8	2
1	4	9	8	6	2	3	5	7
9	2	7	6	1	8	4	3	5
6	3	4	5	9	7	8	2	1
8	5	1	2	3	4	6	7	9
4	9	2	7	8	1	5	6	3
7	1	8	3	5	6	2	9	4
5	6	3	4	2	9	7	1	8

ACKNOWLEDGMENTS

Thanks to my wife and daughter for all the games and puzzles we've played and made together over the years. —**STEVEN CLONTZ**

Toto, thank you for being my best friend—I appreciate the many memories we share, and I look forward to the journey ahead. Jessica, thank you for making my world brighter every day with your adorable smiles and giggles. Grandma, thank you for inspiring me to pursue my passions. Mom, Dad, and Beth, thank you for raising me to be who I am. Last but not least, Ashley and Mariah, thank you for the wonderful adventures. —**JULIE DEMYANOVICH**

ABOUT THE AUTHORS

DR. STEVEN CLONTZ lives with his wife and daughter in Mobile, Alabama. Through his work as a mathematician, professor, and puzzler, Clontz has designed puzzles and games featured by the National Museum of Mathematics and the international puzzle events DASH and Puzzled Pint. He cofounded Mathematical Puzzle Programs, which organizes puzzlehunt events at colleges and universities across the US. His website is Clontz.org.

JULIE DEMYANOVICH is a game developer, publishing games since 2019. From sudoku to point-and-click adventures, Julie orchestrates innovative, engaging experiences for audiences of all ages to enjoy. Outside of serving as executive producer, writer, and artist for a number of published independent games, she is a composer, graphic designer, and video editor. Before developing sudoku puzzles, Julie earned undergraduate degrees in psychology and game development from George Mason University and an MBA from the University of Maryland.

Hi there,

We hope you enjoyed *Sudoku for Brain Fitness*. If you have any questions or concerns about your book, or have received a damaged copy, please contact customerservice@penguinrandomhouse.com. We're here and happy to help.

Also, please consider writing a review on your favorite retailer's website to let others know what you thought of the book!

Sincerely,

The Zeitgeist Team